WILD FLOWERS OF
NAMAQUALAND

A BOTANICAL SOCIETY GUIDE

FOURTH REVISED EDITION

TESSA OLIVER

Annelise le Roux

This guide is a project of CapeNature (Western Cape Nature Conservation Board) and the Botanical Society of South Africa

Published by Struik Nature (an imprint of Penguin Random House South Africa (Pty) Ltd)
Reg. no. 1953/000441/07
The Estuaries No 4, Century Avenue, Oxbow Crescent, Century City, 7441
PO Box 1144, Cape Town, 8000 South Africa

Visit **www.penguinrandomhouse.co.za** and join the Struik Nature Club
for updates, news, events and special offers

First published 1981
Second edition 1988
Third edition 2005
Fourth edition 2015

5 7 9 10 8 6 4

Publishing manager: Pippa Parker
Managing editor: Helen de Villiers
Editor: Emsie du Plessis
Designer: Gillian Black
Proofreader: Tina Mössmer

Reproduction by Hirt & Carter (Pty) Ltd
Printed and bound in China by C&C Offset Printing Co., Ltd

FSC

MIX
Paper from
responsible sources
FSC® C018179
www.fsc.org

Print ISBN: 978-1-77584-131-9 (softcover edition)
Print ISBN: 978-1-77584-320-7 (Collector's edition)
ePub ISBN: 978-1-77584-317-7
ePDF ISBN: 978-1-77584-318-4

DISCLAIMER

Although care has been taken to be as accurate as possible, neither the author nor the publisher makes any expressed or implied representation as to the accuracy of the information contained in this book and cannot be held legally responsible or accept liability for any errors or omissions.

This book contains many quoted examples of medicinal or other uses of plants. The publisher and author do not assume responsibility for any sickness, death or other harmful effects resulting from eating or using any plant described in this book. Readers are strongly advised to consult a health professional, and any experimentation or usage is done entirely at own risk.

LIST OF SUBSCRIBERS

ACKNOWLEDGEMENTS

This has been an exciting project thanks to the many friends, colleagues and Namaqualanders who have played a part in the production of the book. It was certainly very much a combined effort and a privilege to work with such a group of willing and enthusiastic people, all of whom contributed to making this guide a most enjoyable project.

When I think of Namaqualand flowers, I inevitably think of Gretel van Rooyen who freely shares her knowledge of the wonder of survival of the annual species with everybody. She is an integral part of the knowledge and information in this guide and also checked the text from a botanical as well as a language point of view. She and Riaan de Villiers gave us permission to use text and photographs from the *Cederberg Wild Flower Guide* to expand the forth edition of this guide.

Over the last few years I spent special times with the following Namaqualand friends who shared their joy of the veld and their many photographs: Heather Burger, Jean Smith, Lita Cole, Maryna and Helmuth Kohrs, Pieter van Wyk and Veronica Esterhuizen. I would especially like to thank Jean Smith for going through the text with a fine-tooth comb and teaching me a better English.

I would like to thank the Western Cape Nature Conservation Board (CapeNature) for its interest in and continuous support of this project, and the following people for their friendly, invaluable help with various aspects:

■ **Naming of plants** – the staff of the Compton Herbarium (especially Dee Snijman, John Manning, Anthony McGee, Nicola Bergh and Steven Boatwright), the Bolus Herbarium (especially Cornelia Klak), Kirstenbosch National Botanical Garden (Ernst van Jaarsveld), the National Herbarium, as well as the botany departments of the Universities of Stellenbosch, the Western Cape and Pretoria.

■ **Common names and traditional uses of plants** – P. Aggenbag, T.J.N. Beukes, F.J.J. Brand, L. Brandt, C. Brandt, G. & A. Cloete, L. Cole, J.S. Esterhuizen, C. Gagiano, W. Huisamen, J. & K. Joseph, N. Joseph, A.A.J. & M. Kennedy, P. Kotze, H.C. Rossouw, H.L. Schreuder, H. Steenkamp, C. Swart, W.C. van den Heever, C.A. van der Westhuizen, J.C. van der Westhuizen, H. van Zyl and W.P.L. van Zyl; information on palatability of plants for stock – Gerrit Louw.

■ **Comments and suggestions for improvement of the text** – many, many people have read and re-read the texts of all four editions. Without them there would have been no flower guide. First edition: M.F. Thompson, B.J. Ferreira, J.C. Greig, P.H. Lloyd, H. Muller, M.J. Simpson, R.W. Rand, H.A. Scott and E.E. Viviers. Second edition: E.G.H. Oliver, L. Bosman, K.J. Bergh, A. Gillie and M.J. Simpson. Third edition: D. McDonald, I. Rogers, C. Duckitt, M. Botha and D. Lategan. Fourth edition: Gretel van Rooyen, Jean Smith and Kate Steyn.

■ **Photographs** – although most of the photographs were taken by Zelda Wahl, many friends and colleagues have freely supplied photographs for the expanded fourth edition. Photographers are acknowledged alongside the photographs.

Thank you to all who share their awe, passion and knowledge of Namaqualand with me – it remains a joyous privilege. Above all, I thank the Lord for His creation.

CONTENTS

ANNELISE LE ROUX

The true glory of Namaqualand – a profusion of colours from a mixture of annual and perennial plants.

FOREWORD

The Namaqualand is one of the most fascinatingly diverse areas in South Africa and home to thousands of plant species, many of them endemic to this region and many threatened with extinction according to the IUCN's Red-list of species in the area. As the interest in, and knowledge of this diverse flora grows, it has been necessary to repeatedly revise the *Wild Flowers of Namaqualand*, and this is now its fourth edition.

This area is also home to the Knersvlakte Nature Reserve, one of the crown jewels in the country's rich botanical treasure trove and was added to the national network of protected areas in September 2014.

The sheer diversity and high numbers of endemic plant species found in the Namaqualand is what makes it such a vital and popular area for local and international botanists who spend months at a time conducting extensive research here. The plants of course form an integral part of our ecological heritage, especially here in the Western Cape and it is a legacy that I keep close to my heart.

We all have a collective role to play in conserving our environment and thanks to guides such as *Wild Flowers of the Namaqualand*, we can all take responsibility for getting to know more about the biological diversity found in the province. If we do not do it now, it may be too late tomorrow.

The origins of this book date back to 1976 when a discussion at the Clanwilliam Wild Flower show, between the late Professor Ted Schelpe of the University of Cape Town, Mr A.E. Sonntag of the Department of Water Affairs, Forestry and Environmental Conservation, Prof H.B. Rycroft of the National Botanic Gardens and Mrs Kay Bergh of Clanwilliam, resulted in a decision that a guide to these flowers needed to be published. In 1981 the Botanical Society of South Africa published the first *Wild Flower Field Guide, Namaqualand and Clanwilliam*.

However, because both areas are so rich and interesting botanically that they require separate guides, this book is now focused specifically on the Namaqualand. With numerous revisions, it has been carefully edited to feature 868 species.

Compiled by CapeNature and the Botanical Society of South Africa, this guide has been designed to assist students, teachers, tourists, land owners and any other interested parties to appreciate the unique floristic wealth and diversity of this area in South Africa.

This guide is certainly a big collaborative effort. A number of people from Namaqualand have contributed input and photographs and I congratulate in particular Annelise le Roux who has compiled this edition after taking into account the latest taxonomic developments.

I applaud this latest expanded edition of the Namaqualand Guide and recommend it not only to first time buyers but also to owners of earlier editions of the guide.

Dr Razeena Omar
Chief Executive Officer
CapeNature

The thought of spring in Namaqualand generally brings to mind carpets of bright orange, yellow and white flowers. During spring, Namaqualand is a wonderland, unbelievably beautiful in a good year. One stares in amazement, convinced that nothing can be more beautiful, but then a step or so further on, one stops again in awe. This wonder is twofold: that such masses of flowers occur and that each species seems more striking than the next. The flora of Namaqualand is unique. In this region one finds not only carpets of annuals after a good wet winter, but also a wide variety of geophytes, dwarf shrubs and succulents. Geophytes are plants with bulbs, corms or tubers, such as members of the families Hyacinthaceae, pp.119-140, and Iridaceae, pp.42-83. Succulents can vary from a creeping plant like the rank-t'nouroe (*Cephalophyllum rigidum*, p.308) to a large stem-succulent like the botterboom (*Tylecodon paniculatus*, p.156) or a tree-succulent like the quiver tree (*Aloidendron dichotomum*, p.98). This rich heritage of about 3,800 species, of which 28% is endemic, is certainly worth conserving for future generations.

The best flower displays usually occur from early August to mid-September. Weather patterns, particularly hot bergwinds and the timing of the first substantial rainfall, determine both the time and duration of the flower season, and accurate predictions on when and how long a season will last, cannot be made.

With this book, we introduce you to some of the plants of Namaqualand, hoping it may add to your enjoyment of nature. May this joy motivate a desire to conserve and protect this exceptional region, for example by stepping over and not on a plant and leaving litter in the car and not among the flowers.

ZELDA WAHL

Spring in the Namakwaland Klipkoppe. In winter and spring this is a colourful landscape of perennials and many annuals, but in summer it becomes a barren scene.

Summer in the Namakwaland Klipkoppe.

WHAT MAKES THIS FLORA UNIQUE?

According to most definitions, this region is classified as a desert. A barren land for almost three-quarters of the year (summer, autumn and winter), Namaqualand becomes green and colourful for not more than two to three months in spring. This spring-summer transformation is almost unimaginable and only by seeing the change for oneself, can one appreciate it. Namaqualand is unique in being the only desert in the world to have such an extravagant and diverse spring flower display.

Most plants throughout the world require warmth, together with water, to optimally grow, flower and produce seed, usually during spring and summer. They then survive unfavourable growth conditions (low temperatures and water scarcity) normally during the winter months. The climate of Namaqualand (low, sporadic winter rainfall and very hot, dry summers) offers the opposite of optimal growth conditions and is normally not conducive to plant growth. Plants of Namaqualand have adapted and developed many special features to survive cold, relatively moist winters and hot, dry summers. The plants grow slowly and even form flower buds in winter, but only burst into flower in the warmer spring months. They die off and become dormant quickly within two to three weeks at the beginning of summer and survive the very long hot, dry months through different strategies.

Precipitation, not only in amount but also in timing, is the determining factor for growth and good flower displays. Annual rainfall ranges from 50–400mm, averaging 150–200mm. However, owing to the low winter temperatures during the rainy season, little water is lost through evaporation and most of the precipitation in the wet winter months (rain, dew, frost or even snow sometimes) remains available to plant life in the growing season. The fact that it rains in winter is key: summer showers may add considerably to the total annual rainfall, but are lost to plant and animal life because of evaporation on very hot days. The annual rainfall can differ markedly from one year to the next, resulting in a different flower display each year.

Annuals

Annual plants in this region survive the dry period by germinating quickly, and growing, flowering and setting seed during the moist winter and spring. They then survive the dry summer and autumn in the form of seed.

Namaqualand annuals are prolific in their seed production. However, seeds of many species remain dormant for some time, preventing all the seed from germinating after one good shower. This phenomenon of delayed germination ensures that a viable seed store always remains in the soil. Studies have indicated that the overall size of the seed bank in Namaqualand is large; figures of more than 40,000 seeds per square metre have been recorded.

Among the annuals of Namaqualand, a number of species are found that form morphologically distinct types of seeds. Perhaps the best-known example occurs within the genus *Dimorphotheca* (the Namaqualand daisy, pp.433–435), where

The flowers of *Dimorphotheca sinuata* (the real Namaqualand daisy) like all spring flowers, are a source of food for all animals and insects.

the ray-florets produce one type of seed and the disc-florets another. These seeds differ not only with respect to dispersal but also in germination behaviour. Even the plants that grow from the different types of seeds have different traits. Disc-seeds, which are dispersed widely and germinate easily, produce competitively superior plants, whereas ray-seeds, which are not widely dispersed and do not germinate easily, produce plants that are competitively inferior. While disc-plants are opportunistic and are responsible for extending the range of the species, ray-plants have adopted a more cautious strategy and are responsible for maintaining the species in the seed bank, thus ensuring survival of the species under unfavourable conditions.

Perennials

Geophytes survive the summer underground in the form of bulbs, corms or tubers. Some geophytes, such as the suikerkannetjie (*Massonia depressa*, p.130), create their own cool, moist microclimate with their two large, flat leaves on the ground preventing evaporation of ground moisture.

Some evergreen dwarf shrubs have either small leaves like the kapokbos (*Eriocephalus microphyllus*, p.458), limiting the area from which transpiration can occur, or succulent leaves like the vygies or mesems (pp.276–334). Dwarf shrubs with larger leaves like species of *Berkheya* (p.407), shed them during the dry summer months for the same reason. Some plants have green stems (*Euphorbia* species, pp.226–231, and *Hermbstaedtia glauca*, p.262) in which photosynthesis occurs and transpiration from leaves is therefore minimal.

The vygies (pp.276–334), a dominant succulent group in Namaqualand, have many adaptations that prevent water loss. Some that do not have a thick skin, have many large cells filled with water,

Bladder cells on the leaf surface of *Drosanthemum hispidum* seedlings.

called bladder cells. While conditions are moist, these cells are thick and swollen, standing almost upright. The stomata used for carbon dioxide intake and evapotranspiration are in between these cells. When drought sets in, the bladder cells 'wilt', falling flat and covering the stomata, thus preventing water loss through them.

Some vygies, known as stone plants, are very small, have only one or two leaves per plant and never grow much above ground level. These plants are very efficient in recycling their water. When drought sets in, water is withdrawn from the existing leaves and stored in new leaf growth that has begun to erupt in readiness for the subsequent summer, but remains protected underground in the centre of the plant. The existing leaves then dry out and form a white, papery layer that shields the new leaves in the summer months (*Oophytum nanum*, *O. oviforme*, p.330, and *Meyerophytum globosum*, p.326). The white colour also reflects the

rays of the sun and keeps the plant cool. Some vygies, such as the window plants (*Fenestraria rhopalophylla*, p.321) found along the coast of Namaqualand, take evasion even further. They withdraw totally underground in summer and, even in winter when it is wetter, they do not grow above the soil surface. All that is exposed to the air is a small round transparent 'window' that lets in sunlight. The sunlight is used for photosynthesis by the underground sidewalls of the leaves that have chlorophyll.

AREA COVERED BY THIS BOOK

Namaqualand, the land of the Nama people, is situated in the northwestern corner of South Africa. This area was originally known as Little Namaqualand, while southern Namibia (south of Windhoek) constituted Great Namaqualand, now known as Namaland. The region extends from the Orange River southwards to the Olifants River and inland from the coast to just east of Springbok, Gamoep and Kliprand, to the foot of the Bokkeveld Mountains, and then to Vanrhynsdorp and Vredendal. The area covers about 55,000km² in all. Many species, especially annuals, described in this guide are also found in the Clanwilliam area and in Bushmanland, to the east of Namaqualand. However, the Clanwilliam area has a much higher rainfall than Namaqualand, and Bushmanland is a summer-rainfall area, resulting in different perennial vegetation. These two regions are therefore not discussed in this guide.

Namaqualand can be divided into four regions according to the features of the physical environment, for example, the topography, soils, geology and climate, coupled with the vegetation and animal life found in these habitats. These regions are the Richtersveld, the Namakwaland Klipkoppe, the Coastal Plain of Namaqualand and the Knersvlakte.

Meyerophytum globosum in summer with the previous season's old leaves that have withered completely into a white papery covering, protecting the round young buds of the next season.

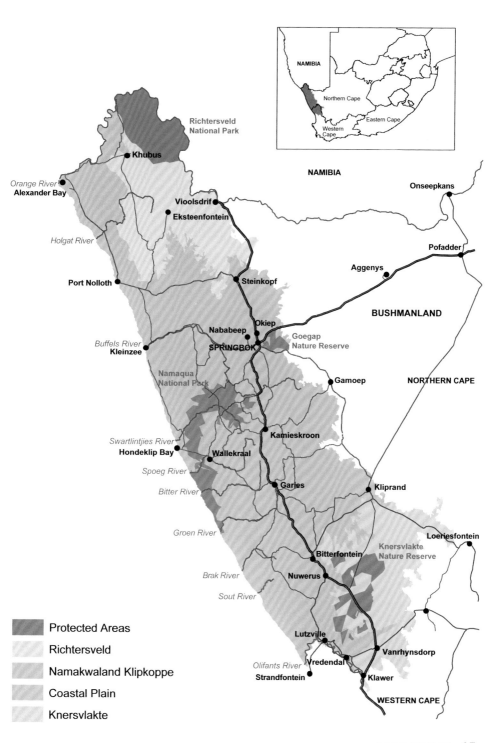

NAMIBIA

Northern Cape

Eastern Cape

Western Cape

Richtersveld
National Park

● Khubus

NAMIBIA

● Onseepkans

Orange River
Alexander Bay ●

● **Vioolsdrif**
● **Eksteenfontein**

● Pofadder

Holgat River

Aggenys ●

● **Port Nolloth**

● **Steinkopf**

BUSHMANLAND

● Okiep
Nababeep
SPRINGBOK

Goegap
Nature Reserve

Buffels River
Kleinzee ●

*Namaqua
National Park*

● Gamoep

NORTHERN CAPE

● **Kamieskroon**

Swartlintjies River
Hondeklip Bay ●

● **Wallekraal**

Spoeg River

● Garies

Bitter River

● **Kliprand**

Groen River

Loeriesfontein ●

● **Bitterfontein**

Knersvlakte
Nature Reserve

Brak River

● **Nuwerus**

Sout River

● **Lutzville**

Vanrhynsdorp ●

Olifants River
Strandfontein

● **Vredendal**

● **Klawer**

WESTERN CAPE

Protected Areas

Richtersveld

Namakwaland Klipkoppe

Coastal Plain

Knersvlakte

The Richtersveld, mountainous and barren but with many endemic plants.

ZELDA WAHL

The Richtersveld

The Richtersveld occupies the very hilly and mountainous extreme northern corner of Namaqualand. There are many different demarcations of the Richtersveld, ranging from the broad definition used in the previous editions of this book (the Orange River and the Atlantic Ocean forming the northern and western boundaries respectively, and Port Nolloth, Steinkopf and Vioolsdrif being the southern and eastern limits), to much narrower definitions, such as that pertaining to the area occupied by the local community of the Richtersveld. However, in this edition of the guide, the name Richtersveld refers to a biogeographical area, meaning that the environment, plant and animal life, as well as the climate, are closely related. This area, indicated on the map on p.15,

is a very mountainous region with many geological formations, volcanic and sedimentary in origin, consisting mainly of granite, gneiss, limestone, schist, layered shale and scatterings of white quartz. Erosion through centuries has created harsh, rugged, yet colourful and beautiful scenery.

On average, the vegetation is less than 50cm high, sparse and known for the large variety of succulent species and many growth forms. These succulents belong mainly to the vygie group (Aizoaceae, pp.276–334), the *Crassula* group (Crassulaceae, pp.156–171) and the *Pelargonium* group (Geraniaceae, pp.172–185). The area is used for extensive small-stock farming, mainly goats. The Richtersveld is named after a missionary who served in the area in the 1900s.

The Coastal Plain

The Coastal Plain of Namaqualand, sometimes also called the Sandveld, is a strip about 30km wide that runs along the coast from the Orange River to the Olifants River. It comprises loose, white sand near the coast and deep red sand further inland. In the interior of the Coastal Plain, the soil is a more calcareous, red loam in a slightly undulating landscape. Rainfall is less than 50mm per annum from the Orange River to Port Nolloth, 50–100mm per annum southwards to the Groen River and 100–150mm per annum southwards to the Olifants River.

At the coast the vegetation is usually low, about 30cm high, and geldjiesbos (*Roepera cordifolia*, p.186), several species of *Drosanthemum*, a gazania (*Gazania splendidissima*, p.412) and karoorapuis (*Crassothonna sedifolia*, p.427) are generally found here. Only a little further inland the vegetation is on average 1m high, with t'arra-t'kooi (*Stoeberia utilis*, p.335) the most common vygie bush.

A kilometre or so further inland, the skilpadbos (*Roepera morgsana*, p.187), ossierapuis (*Crassothonna cylindrica*, p.426) and other shrubby vygies, especially *Lampranthus stipulaceus* (p.324) and volstruisvygie (*Jordaaniella spongiosa*, p.323), are common. Extensive small-stock farming is practised here, but mining also contributes to the economy of this region.

The Namakwaland Klipkoppe

This region, sometimes referred to as Namaqualand Hardeveld, is considered the true Namaqualand by the local inhabitants and is characterised by distinctive round, rocky granite hills separated by sandy plains. Many of these rocky hills have conspicuous large, flat, sometimes rounded, and exposed rock surfaces. The region can be regarded as an escarpment, about 50km wide, between the low-lying Coastal Plain and the Bushmanland plateau, and stretches from Steinkopf to Nuwerus and southwards to just north of Lutzville. The rainfall varies from 100 to

The Coastal Plain of Namaqualand is home to many low, flowering shrubs.

ZELDA WAHL

Namakwaland Klipkoppe with typical sandy plains and granite hills.

200mm per annum but at some places in the Kamiesberg up to 400mm per annum has been recorded.

The vegetation is 0.5–1m high and slightly higher in the rocky hills than on the plains. One of the most common dwarf shrubs found on the plains is kraalbos (*Galenia africana*, p.282). It may be considered a pioneer plant as it is the first perennial to appear on abandoned fields and the only one remaining on very badly overgrazed areas, as small-stock do not utilise it. Other dominant species, especially in the hills, are kapokbos (*Eriocephalus microphyllus*, p.458), skaapbos (*Osteospermum sinuatum* and *O. oppositifolium*, p.432), perdebos (*Didelta spinosa*, p.408), skilpadbos (*Roepera foetida*, p.186), fluitjiesbos (*Calobota sericea*, p.197) as well as many vygie dwarf shrubs of the Aizoaceae such as swart t'nouroebos (*Ruschia robusta*, p.333), langbeen t'nouroebos (*Leipoldtia schultzei*, p.325) and fyn t'nouroebos (*Drosanthemum hispidum*, p.318). A few ephemerals that are usually obvious and abundant are pietsnot (*Grielum*

humifusum, p.252), dassiegousblomme (*Osteospermum amplectens* and *O. hyoseroides*, p.430), Namaqualand daisy (*Dimorphotheca sinuata*, p.435), hongerblom (*Senecio arenarius*, p.416), teebossie (*Leysera tenella*, p.436) and many sporrie species (*Heliophila* species, pp.233–237). The most important farming activity is small-stock farming, especially karakul sheep, but wheat is also grown on the plains, mainly in the Kamiesberg where the rainfall is slightly higher.

It is on these wheat fields that are not cultivated every year, or have not been ploughed for a few years, that a carpet of flowers springs up – very much like a crop of wheat – usually of one species and one colour. In the natural undisturbed veld itself, however, there is a glorious spectacle of many species and colours of annual flowers, mixed in between the flowering perennials. It is this diverse display of multiple species of annuals and perennials, all in flower at the same time, that reflects the true glory of Namaqualand, not monocultures of species on disturbed areas.

HEUWELTJIEVELD

A characteristic of the gently undulating landscapes in all four regions of Namaqualand (and of other arid Karoo areas), is the presence of regularly spaced circular patches or mounds, 10–15m in diameter, known as 'heuweltjies'. There is much speculation about their origin, but the most acceptable theory seems to be that they are extinct termite mounds, although there may still be termite activity in a few. Termites (*Microhodotermes viator*) cause hardening and other changes in the soil composition, and create a mound, which also serves as an ideal habitat for burrowing animals such as porcupine, aardvark and Brand's whistling rats. Plant species found on these mounds differ considerably from those in the areas between the mounds. The soil structure of the mounds is sensitive to disturbance, and trampling by grazing animals and burrowing activity can lead to the denudation of the 'heuweltjie'.

ANNELISE LE ROUX

Heuweltjieveld near Nuwerus with typical disturbed vegetation on the 'heuweltjie' (small hill) in the foreground of the picture.

The Knersvlakte

The Knersvlakte lies from north to south between Bitterfontein and Vanrhynsdorp, and from east to west between the foot of the Bokkeveld Mountains near Nieuwoudtville and the Coastal Plain of Namaqualand. The rainfall here varies from 100 to 200mm per annum and the vegetation is low, 10–50cm high. The landscape is typified by very low, rolling hills covered in small, white quartz pebbles and with very saline soils. In these quartz patches there are unplant-like succulents such as bababoudjies and vingervygies (*Argyroderma* species, pp.304–305), krapogies (*Oophytum oviforme*, p.330) and duim-en-vinger (*Mesembryanthemum digitatum* subsp. *digitatum*, p.290). Many other members of the mesem group are found on the Knersvlakte, with the odd Asteraceae shrub. Each quartz patch may have its own species composition, and even the size of the white quartz pebbles determines the species that are present. Extensive small-stock farming is practised and mineral deposits are also found.

ZELDA WAHL

ZELDA WAHL

Above: White quartz pebble patches are interspersed with small succulents on the Knersvlakte.

Left: White quartz pebbles reflect the sun, creating a slightly cooler environment for succulents such as *Argyroderma delaetii* and *Oophytum oviforme*.

SPECIES ILLUSTRATED AND DESCRIBED

The families and genera in this guide are arranged according to a botanical classification system that reflects new knowledge about relationships between the various plant groups. This means that similar-looking plants are grouped together in the guide. Species appear alphabetically under their genus, except in the case of *Mesembryanthemum* and *Osteospermum* where they are presented in groups of related species.

The variety of species in Namaqualand is so large that it is impossible to illustrate and describe them all in one book. As most species flower in spring – the time of year when most people visit Namaqualand – many of the plants that are common

Various explanations for the origin of the name 'Knersvlakte' exist. Some say it was a case of 'tandekners' (gnashing of teeth) for those travelling through the area by ox-wagon in days gone by. According to others, the original name 'Knechtsvlakte' (servant's plain) was changed to 'Knersvlakte'. This plain was rented from the state in the early days and a tenant was called a 'knecht van de staat' (servant of the state).

and conspicuous during this season are featured. Annual species likely to appear every year, especially the most common ones, have been included in the book. Some species that are very similar in appearance to those illustrated have also been included to help with identification.

Many species found in Namaqualand belong to only a few families and genera and, because they look very much alike, can usually be identified to genus level by finding similar-looking species in the guide. However, visitors to the area will also encounter species that are not illustrated and do not even look like any of those in the guide, simply because of limited space in the book.

Species endemic to Namaqualand are indicated as such in the boxes with the names of the plants. Some endemics are common in Namaqualand while others are more rare and localised. Not all species illustrated are indigenous. Some alien species that have become naturalised and are commonly found in the veld are illustrated and the country of origin is given.

WHAT TO LOOK FOR WHEN IDENTIFYING A SPECIES

In this book the photographs are the primary guide. With the plant in front of you, look at the images, decide which species it most closely resembles and then read the text, comparing your plant for size, shape of the leaves, flowers and other details. Maximum dimensions are generally given, but bear in mind that specimens are usually smaller in years of low rainfall. The distinguishing features of superficially similar species are also given. In certain families there are specific characters that will facilitate the identification of the species, for example the 'pods' of the sporries (Brassicaceae, pp.233–237), the dry fruits of the vygies (pp.276–334) and the involucral bracts and seeds in the daisy family (Asteraceae, pp.398–465).

COMMON NAMES AND USES OF THE PLANTS

The common names in English, Afrikaans and Nama were obtained from Namaqualanders themselves. Keep in mind that different names are sometimes given to the same species in different regions and that not all regional names appear in this wild flower guide.

Some interesting facts and traditional uses of some of the plant species illustrated are also mentioned.

Palatability classes for stock are given in this guide as: highly palatable (animals show a preference to these species and will select them first), palatable (readily eaten by animals) and unpalatable (plants that are only eaten if there are no palatable plants available). In general, annuals can be considered as highly palatable species to which animals are very partial. They are not only herbaceous, soft and easily digested, but the flowers and seeds are highly nutritious because many of the nutrients needed by the seedlings are stored there. Even dry, dead annuals are very palatable. Selective grazing of flowers is sometimes evident on road verges. Plants between a road and a fence are not accessible to animals for grazing and roadsides often produce some of the best flower displays. Since annuals, as well as the soft, deciduous leaves, flowers and seeds of perennials, are not available during the summer months, animals are then forced to live off unpalatable plants, especially dwarf vygie shrubs or mesems (pp.276–334), which have evergreen, succulent leaves.

BOTANICAL NAME CHANGES

Over the years, scientists have attempted to standardise biological names worldwide. This has led to what is called a binomial system, with a scientific name consisting of a genus name and a species name, for example *Ferraria divaricata*. The genus name indicates the relationship with other plants, and therefore shows that

Ferraria divaricata is closely related, and has certain similarities in appearance, to *Ferraria ferrariola* and *Ferraria schaeferi*. Since there are so many different plants and not sufficient knowledge about all of them yet, botanists are still finding through research that certain species were incorrectly classified in the past. As a result, some names have had to be changed according to the international rules of nomenclature. It is therefore inevitable that with the revision of a plant book such as this, there will be some changes in the botanical names. Other than changes in genus and species names, plants are sometimes also reclassified under a different family. For example, recent genetic studies indicate that some genera previously placed under the family Mesembryanthemaceae show very close relationships with members of the Aizoaceae and these are therefore now classified under the latter family. Where a name has changed since the previous edition of this guide, the previous name (synonym) is now given in brackets after the current name. It is also inevitable that plants will be misidentified, especially in a region like Namaqualand with so many species and so few botanists studying these plants. Misidentifications in the third edition are also indicated in brackets after the now correct name.

CONSERVING OUR HERITAGE

Many plant species of Namaqualand are endemic and are found nowhere else in the world. Endemics are usually species that have evolved to survive under very harsh conditions and are therefore very specialised. This means that any type of change in their surroundings would affect them, mostly adversely. Larger bushes usually protect and act as 'nurse plants' for the germination and survival of seedlings. The thinning of vegetation cover through grazing will result in the disappearance of plants because their protection and shade has diminished. Indiscriminate grazing at the wrong time of the year will result in total disappearance of seedlings as they are the most sought after plants in a desert environment. Also, stone plants may not be feed for animals, but trampling of these plants can eradicate them.

Human curiosity about the unusual and our possessiveness is another threat to our unique succulent flora. Most small succulent plants reach an age of up to 50 years and recruitment may take place only once or twice in the lifetime of such a plant. In addition, these plants are usually restricted to small areas, making it easy for a collector to remove all the plants of one population at a time. Removal of plants from the veld can lead to extinction of a species. Please buy plants for your home or garden from a reputable nursery where the plants have been cultivated from seeds, not transplanted from the veld.

Although there are only a few alien invader plants in this region, they should still be eradicated as any threat in a desert area is a major one. Invaders include Jantwak (*Nicotiana glauca*, p.360) and the Mexican poppy (*Argemone ochroleuca*, p.153), and are usually found in disturbed areas, often on abandoned fields.

An abundance of one of the following indigenous species is an indication of over-utilisation of the vegetation and such areas should receive special care: kraalbos (*Galenia africana*, p.282), asbos (*Mesembryanthemum subnodosum*, p.296) and soutslaai (*Mesembryanthemum guerichianum*, p.285). Grazing should be kept to a minimum to encourage recovery of the natural diversity of the vegetation.

Nature gives abundantly of its beauty but this should not be abused. In Namaqualand, the profusion of flowers is nature's way of ensuring the survival of species adapted to the extreme climatic conditions. One flower may produce a hundred seeds, of which two, maybe

Road verges protected from grazing produce some of the best flower displays.

ANNELISE LE ROUX

three, will survive a four- or five-year drought period. One of our three seeds may become food for an ant and another could fall in a place where it may not be able to germinate. This leaves only one seed to germinate and take part in the mass display of flowers in the next good season. Animals only destroy plants in order to survive, but humans often destroy without thought or reason. Trampling, littering and vehicle tracks in the veld steal some of the magic of carpets of flowers from other people. Please help the farmers to conserve this natural beauty. The conservation of nature is for the benefit of humankind and can be brought about only by us.

Apart from the beauty of nature, nature conservation is also essential for our survival. An insignificant little plant that is ploughed up or bulldozed, could possibly have been a cure for cancer, or a hybrid of such a veld plant could perhaps have been developed as an important cultivated source of food.

Protected areas

Each area in Namaqualand has its unique scenery and plant species. It is therefore essential that a network of reserves protects this heritage. There are a number of areas with different types of conservation status – national parks, provincial nature reserves and local nature reserves, each not only contributing to the preservation of species, but providing opportunities for ecotourism, recreation and education.

The different vegetation types and habitats of the various geographical regions in Namaqualand are represented in present conservation areas such as the Richtersveld National Park, the Goegap Provincial Nature Reserve, the Namaqua National Park and the Knersvlakte Nature Reserve. These areas are there for the public to enjoy and support. Please keep to the rules and regulations of the different areas; they are there to ensure that all visitors enjoy their experience, now and in the future.

SPECIES DESCRIPTIONS

ZELDA WAHL

Wurmbea stricta

(formerly *Onixotis triquetra*)
waterblom, vleiblommetjie
COLCHICACEAE

A perennial herb up to 45cm high, with a corm. The 3 leaves are triangular, smooth and up to 40cm long, tapering towards the tip and with a sheathing base; the 2 upper leaves are just below the flowers and the third leaf is at the base of the plant. White to pink to pale purple flowers up to 1.5cm in diameter are borne close together on a long stalk.

Found in areas with seasonal waters in the Kamiesberg and also southwards to the Cape Peninsula and eastwards to Port Elizabeth.
Flowers August to September, but occasionally from May to November.

NOEL VAN ROOYEN

Wurmbea variabilis

COLCHICACEAE

A perennial herb up to 20cm high, with a corm. The 3 leaves are egg- to broadly sword-shaped, erect or recurved. The foul-scented flowers are stalkless; the petals are greenish to cream-coloured with a maroon median spot, sometimes with maroon margins, and are united at the base to form a tube; the filaments are 3–5mm long; the styles are free and awl-shaped.

Found in moist, sandy or loamy, often stony soil in relatively open patches of low, shrubby vegetation in the Kamiesberg and also in other mountainous areas of the Northern, Western and Eastern Cape.
Flowers August to October.

There are eight species of *Ornithoglossum* in southern Africa of which five occur in Namaqualand.

Ornithoglossum undulatum
spinnekopblom

COLCHICACEAE

A perennial herb up to 20cm high, with a corm. The 2–4 sword-shaped leaves are greyish-green; the margin is smooth or undulate. The fragrant, nodding, asymmetrical flowers have recurved petals that are white to pink with purple or maroon tips, sometimes with a reddish blotch near the base; the filaments are needle-like or slightly swollen and curve downwards.

Found on rocky slopes or stony flats from Namaqualand to the Tankwa Karoo and the Eastern Cape.
Flowers April to July.

JEAN SMITH

Ornithoglossum vulgare
spinnekopblom

COLCHICACEAE

A perennial herb up to 15cm high, with a corm. The grey-green, sword-shaped leaves are spreading to recurved, variable in length, with a sheathing base; the margin is entire or rarely slightly undulate. Nodding flowers, with reflexed greenish petals with maroon or purplish-brown margins, are borne on long stalks; the filaments are thickened or swollen at the base. Poisonous to stock.

Found on stony or gravelly flats throughout Namaqualand and also elsewhere in South Africa and into tropical Africa.
Flowers August to October.

TESSA OLIVER

Ornithoglossum zeyheri
spinnekopblom

COLCHICACEAE   endemic

A perennial herb up to 5cm high, with a corm. The suberect, greyish, linear, sword-shaped leaves are congested at the base; the margin is very undulate. The flowers are erect, pale greenish and flushed purple, with slightly swollen sections halfway up the filaments.

    Found in gravelly pockets on granite or sandy flats throughout Namaqualand. **Flowers May to July.**

COLCHICUM
There are about 50 species of *Colchicum* in southern Africa of which 17 occur in Namaqualand.

Colchicum albofenestratum

COLCHICACEAE   endemic

A perennial herb growing flat on the ground, with a corm. The 2 lower leaves are grey-green, flat and sword- to egg-shaped; the margin is crisped. The uppermost, erect, bract-like leaves are thinly textured and white, with conspicuous green longitudinal and transverse veins; the margin is entire or slightly undulate. The 1 or 2 flowers have filaments that are as long as the petals, usually 7mm, and the oblong anthers are 5–6mm long.

    Found on dolomite hills on the Knersvlakte. **Flowers July to August.**

Colchicum capense
subsp. *ciliolatum*
(formerly *Androcymbium ciliolatum*)
uilblaar, kokerdoosblom
COLCHICACEAE

A perennial herb up to 5cm high, with a deep-seated corm. The 2 sword-shaped, basal leaves, borne at ground level, are up to 15cm long; the margin is minutely fringed. The egg-shaped, erect to spreading bracts are white, or sometimes pale green, and surround a cluster of many small flowers; the green or purple filaments are 6–8mm long and the anthers about 2mm long.

Found in sandy, often moist places throughout Namaqualand, southwards to Piketberg and eastwards to the Roggeveld Mountains and the Tankwa Karoo. **Flowers July to August.**

VERONICA ESTERHUIZEN

Colchicum circinatum
COLCHICACEAE   endemic

A perennial herb up to 5cm high, with a corm. The lower, often purple-spotted leaves are narrowly coiled and frequently undulate; the undersurface is sometimes hairy; the margin is smooth or hairy. The uppermost, erect, bract-like leaves are egg-shaped, blue-green and often purple-spotted; the margin is smooth or hairy. The green flowers are 5–6mm long; the filaments are 6–8mm long, with the anthers 4–5mm long.

Found on stony flats or slopes throughout Namaqualand. **Flowers July to September.**

HEATHER BURGER

JEAN SMITH

Colchicum cruciatum
COLCHICACEAE   endemic

A perennial herb up to 1.5cm high, with
a corm. The 3–5 spreading, oblong,
sword-shaped leaves are about 2.5cm
long and have very small papillae on the
upper surface; the margin is minutely
toothed. The single flower is white with
pinkish or green tips and the filaments
are thread-like.

Found on coarse, sandy flats in the
Namakwaland Klipkoppe.
Flowers in July.

GRETEL VAN ROOYEN

Colchicum dregei
COLCHICACEAE

A perennial herb up to 5cm high, with a
corm. The 3 linear, sword-shaped leaves
are 1–10cm long and have a sheathing
base; the margin is minutely toothed. The
1 or 2 flowers have green petals that are
4–8mm long; the filaments widen into
purple glands at the tips, just below
the anthers.

Found on rocky slopes, usually in the
shade of boulders, in the Namakwaland
Klipkoppe and elsewhere in the
Western Cape.
Flowers June to August.

Colchicum scabromarginatum
(formerly *Androcymbium scabromarginatum*)
skuite
COLCHICACEAE   endemic

A perennial herb up to 5cm high, with a
corm. The 2 lower basal leaves are stiff,
leathery and egg-shaped to oblong; the
margin is smooth or minutely hairy. The
third leaf is similar, but shorter, paler
green, more upright and egg-shaped; it
folds over a cluster of many flowers. The
flowers are honey-scented, with filaments
5–9mm long and anthers 4–5mm long.
     Found in sandy soil in the Richtersveld
and Namakwaland Klipkoppe.
Flowers August to September.

Colchicum albomarginatum is very similar
in appearance, but has sword-shaped
leaves and thickened, white leaf margins.

CAROLIN MAYER

Colchicum villosum
COLCHICACEAE   endemic

A perennial herb up to 5cm high, with a
corm. The lower narrowly inversely egg-
shaped leaves are about 5cm long and
have long, shaggy hairs on the upper
and lower leaf surfaces. The slightly
erect, narrowly egg-shaped bracts have
slightly shorter hairs and surround a
single, pale green flower; the light
brown filaments are 8mm long, with
anthers about 4mm long.
     Found on gravelly slopes among rock
ledges in the Namakwaland Klipkoppe.
Flowers July to August.

HELGA VAN DER MERWE

DISA
There are about 150 species of *Disa* in southern Africa of which four occur in Namaqualand.

Disa karooica
ORCHIDACEAE

A robust perennial herb up to 60cm high, with a tuber. The basal, strap-shaped to linear leaves are usually withered or dry at flowering time. Few to many cream-coloured flowers, with purple markings and a single spur 3–5cm long, are borne on a stalk.

Found on dry slopes or in leaf litter under shrubs in the Kamiesberg and also in the Roggeveld Mountains towards Sutherland.
Flowers October to November.

RUTH COZIEN

Disa macrostachya
ORCHIDACEAE   endemic

A deciduous, robust perennial herb up to 35cm high, with a tuber. The 4–8 strap-shaped leaves are clustered at the base of the flowering stalk. Flowers with a single spur 6–8mm long are borne on the sparsely branched stalk; the median sepal hood is maroon-purple and the lateral sepals are green or white with sparse maroon mottling; the petals and lip are lime-green.

Found in damp, sandy, stony soil near streams in the Kamiesberg.
Flowers in September.

Disa spathulata
begging hand, oupa-en-sy-pyp
ORCHIDACEAE

A deciduous perennial herb up to 30cm high, with a tuber. The leaves are flat and linear. The flowers are 2-lipped: the upper lip is purplish and helmet-shaped with a short spur; the lower lip is yellowish-green and spoon-shaped, with a long limb and blades deeply 3-lobed with wavy margins.

Found in damp areas, but also on well-drained soil in shale in full sun in the Namakwaland Klipkoppe and southwards to Caledon and Uniondale.
Flowers September to October.

ANNELIS LE ROUX

PTERYGODIUM
There are about 20 species of *Pterygodium* in southern Africa of which five occur in Namaqualand.

Pterygodium catholicum
moederkappie
ORCHIDACEAE

A deciduous perennial herb up to 25cm high, with a tuber. The 2–4 grey-green leaves are borne at the base of the flowering stalk, which is often tinged red. The few greenish-yellow flowers become orange-red with age.

Found in moist soil in the Renosterveld in the Kamiesberg of Namaqualand and the mountains of the Bokkeveld, extending to Port Elizabeth.
Flowers September to November.

ANNELIS LE ROUX

Pterygodium crispum
(formerly *Corycium crispum*)
ORCHIDACEAE

A deciduous perennial herb up to 30cm high, with a tuber. The leaves are numerous and sword-shaped, up to 12cm long and 2cm across, with dark purplish spots at the base; the margin is crisped. Yellow and green flowers, about 8mm long and with green lips, are borne close together on a leafy stalk.

Found in sandy soil in the Namakwaland Klipkoppe, on the Knersvlakte and also southeast to the Roggeveld Mountains and Albertinia. **Flowers September to October.**

ZELDA WAHL & COLLA SWART

Pterygodium hallii
ORCHIDACEAE

A deciduous perennial herb up to 50cm high, with a tuber. The sword-shaped leaves overlap on the flowering stalk. Numerous green flowers are borne close together on the stalk, the petals facing down; the central lip appendage is a paler green and the uppermost lip has dense, dark purple spots.

Found on damp, open ground among shrubs in the Namakwaland Klipkoppe and also in other mountainous parts of the Karoo and the Western Cape. **Flowers August to October.**

ANNELISE LE ROUX

Holothrix aspera
ORCHIDACEAE

A deciduous perennial herb up to 10cm high, with a tuber. The 2 opposite, hairless leaves are broad, egg-shaped, up to 3.5cm long and 3cm across, entire and flat on the ground. White flowers with purple markings are borne loosely on a leafless stalk; they are up to 1cm long and distinctly hooded, with a lower, hand-like, divided lip protruding forward; the 7 lobes are broad and flat and the spur is 2–5mm long.

Found on moist mountain slopes in Namaqualand and in the rest of the Western Cape.
Flowers June to October.

ZELDA WAHL & RUPERT KOOPMAN

Holothrix filicornis
ORCHIDACEAE

A deciduous perennial herb up to 25cm high, with a tuber. The 2 basal, egg-shaped leaves are up to 5cm across and hairless. There are many flowers on a stalk, each one ending in 11 long, fine, spreading, white petals and with a prominent spur up to 1cm long.

Found in rock crevices and on rocky slopes in Namaqualand and Namibia.
Flowers May to July.

HEATHER BURGER

ORCHIDACEAE **35**

SATYRIUM
There are about 40 species of *Satyrium* in southern
Africa of which five occur in Namaqualand.

Satyrium erectum
ewwa-trewwa
ORCHIDACEAE

A deciduous perennial herb up to 40cm
high, with a tuber. The 2 opposite leaves are
broad, egg-shaped, up to 15cm long and
10cm across, entire and flat on the ground.
Pale to deep pink flowers, with darker
markings and 2 spurs up to 1cm long, are
borne loosely on a sheathed stalk.
    Found among rocks in the
Spektakelberg and the Kamiesberg in
Namaqualand, also in the Western and
Eastern Cape.
Flowers July to October.

HEATHER BURGER

Satyrium pumilum
aasblom
ORCHIDACEAE

A deciduous perennial herb up to 5cm
high, with a tuber. Forms a low rosette of
leaves on the ground. The flowers nestle
among leaves; the outside of the flower
is green and the inside a dull greenish-
yellow marked with transverse bars of
dark maroon; the petals are fused for
most of their length and the flowers emit
a fetid odour.
    Found in sandy, damp, open areas
between shrubs in the Kamiesberg
and southwards to the Western Cape,
extending to Riversdale.
Flowers September to November.

ANNELISE LE ROUX

PAURIDIA
There are 26 species of *Pauridia* in southern Africa of which seven occur in Namaqualand.

Pauridia aquatica
(formerly *Spiloxene aquatica*)
vleiblommetjie
HYPOXIDACEAE

An aquatic perennial herb up to 35cm high, with a corm. The 2–6 erect leaves are narrow, cylindrical, channelled and up to 35cm long. The 5 or 6 flowers are white with a green underside and are up to 2cm in diameter.

Found in seasonal ponds or streams in Namaqualand, from Steinkopf to the Kamiesberg and also southwards to the Cape Peninsula.
Flowers June to November.

RIAAN DE VILLIERS

Pauridia scullyi
(formerly *Spiloxene scullyi*)
sterretjie
HYPOXIDACEAE

A deciduous perennial herb up to 20cm high, with a corm covered with soft, fine fibres. The numerous spreading leaves are narrow, linear, hairless and V-shaped in cross-section, up to 15cm long and 6mm across. Yellow flowers about 2cm in diameter are borne on long stalks.

Found on the rocky hills of the Richtersveld and Namakwaland Klipkoppe, usually in crevices of big rock faces on south-facing slopes, also in Namibia.
Flowers July to October.

HEATHER BURGER

OLGA CRUZ

Pauridia serrata
(formerly *Spiloxene serrata*)
sterretjie
HYPOXIDACEAE

A deciduous perennial herb up to 20cm high, with a corm. The 6–13 linear, channelled, suberect leaves are about 1.5mm wide; the margin is obscurely and minutely toothed. The flowers are yellow or white, rarely orange, with a green underside.

Found in damp places on flats, often around boulders, in Namaqualand and southwards to the Cape Peninsula and Worcester.
Flowers July to September.

MARYNA KOHRS

EMPODIUM
There are nine species of *Empodium* in southern Africa of which three occur in Namaqualand.

Empodium flexile
HYPOXIDACEAE

A deciduous perennial herb up to 10cm high, with a corm. The narrowly sword-shaped, pleated leaves have fringed ribs and are dry (the previous year's dead leaves) or emerge only at flowering. The petals of the yellow, sweetly scented flowers are outspread, but only slightly spreading near the base; the long, spreading anthers are tipped with fleshy, sterile appendages.

Found on stony flats in the Namakwaland Klipkoppe, on the Knersvlakte and also elsewhere in the Karoo.
Flowers April to June.

Empodium namaquensis
Namakwa-sterretjie
HYPOXIDACEAE

A deciduous perennial herb up to 30cm high, with a corm. The 1–3 broadly sword-shaped leaves are pleated lengthwise, up to 4.5cm long and 2–3cm wide; the margin is shortly fringed. Sweetly lemon-scented flowers are bright yellow inside and greenish outside; the anthers are erect, with no appendages. The flowers appear before the leaves and open in full sun daily between 10:00 and 15:00.

Found in sandy soil on rocky slopes from Namaqualand to Graafwater in the Western Cape.
Flowers April to June.

VERONICA ESTERHUIZEN

WALLERIA
There are two species of *Walleria* in southern Africa of which one occurs in Namaqualand.

Walleria gracilis
TECOPHILAEACEAE

A slender, branched climber up to 50cm long, with a corm. The weak stem has many hooked prickles. The linear leaves are curved or spirally twisted at the tips. Sweetly scented, drooping flowers are white with a purple centre; the yellow anthers have a purple band halfway up their length and are fused at the tips.

Found in sandy soil among rocks in the Richtersveld, and in the Giftberg and Cederberg in the Western Cape.
Flowers June to July.

JOHN MANNING

CYANELLA
There are nine species of *Cyanella* in southern Africa of which seven occur in Namaqualand.

Cyanella hyacinthoides
blouraaptol
TECOPHILAEACEAE

A deciduous perennial herb up to 30cm high, with a corm. The leaves, up to 20cm long, are very variable, ranging from suberect and narrow to flatter, broader and sometimes crisped. Bluish-mauve, sometimes white flowers, about 2cm in diameter and often veined, are borne on a sparsely branched stalk; the arched filaments of the 5 upper stamens are fused for part of their length and the anthers are tubular, while the sixth lower stamen is straight with an oblong anther.

Found in rocky places throughout Namaqualand and southwards to the Western Cape.
Flowers August to November.

Cyanella lutea
geelraaptol
TECOPHILAEACEAE

A deciduous perennial herb up to 35cm high, with a corm. The tufted, basal, linear leaves are hairless, soft and pliable and slightly ribbed; the margin is entire. Sweetly scented yellow flowers, often with maroon veins and occasionally spotted, are borne on a much-branched stalk; the anthers are yellow and frequently spotted with maroon or black and the filaments of the 5 upper stamens are free and the anthers tubular, while the sixth lower filament has a spoon-shaped anther.

Found mostly in loamy soil in the Namakwaland Klipkoppe, also in the Bokkeveld and Roggeveld Mountains and throughout the Western and Eastern Cape.
Flowers September to October.

Cyanella orchidiformis
wilderaap, raaptol
TECOPHILAEACEAE

A deciduous perennial herb up to 40cm high, with a corm. There are usually 5 leaves in a basal tuft or rosette, sword-shaped and up to 13cm long and 2.5cm across. Mauve flowers, 2–2.5cm in diameter and with purple markings, are borne loosely on a sparsely branched stalk; the 6 anthers are yellow and grey, with the 3 upper anthers arched, the median anther longer than the lateral ones and the 3 lower anthers straight. In the past, the corms were often eaten by the people of Namaqualand.

Found in rocky places throughout Namaqualand and southwards to Clanwilliam, also in Namibia.
Flowers July to September.

HEATHER BURGER

Cyanella ramosissima
TECOPHILAEACEAE

A deciduous perennial herb up to 20cm high, with a corm. The linear, channelled, hairless leaves have distinct venation and are borne in a basal tuft; the margin is entire to finely toothed. Fragrant pink to mauve flowers with darker veins are borne on a suberect stalk; there are 6 anthers, the median anther of the upper 3 larger than the lateral ones; style much longer than the 3 lower stamens.

Found on stony flats in the Richtersveld and also in Namibia.
Flowers July to September.

PIETER VAN WYK

MARIUS WHEELER

LAPEIROUSIA
There are about 40 species of *Lapeirousia* in southern Africa of which 17 occur in Namaqualand.

Lapeirousia angustifolia
(previously misidentified as *Lapeirousia pyramidalis*)
IRIDACEAE   endemic

A deciduous perennial, tufted herb up to 15cm high, with a corm. The few linear to sword-shaped leaves are ribbed, the lowermost one the longest, often exceeding the stem. The flowers are borne between green, channelled or folded bracts that have a shallow notch on a rounded tip. The 6–10 sweetly scented flowers are whitish to pink or pale blue, sometimes flushed purple, the slightly smaller, lowermost petals with a purple mark pointing downwards.

Found in sandy or stony soil, sometimes in quartz pebbles, on the Knersvlakte. **Flowers July to September.**

ZELDA WAHL

Lapeirousia arenicola
IRIDACEAE   endemic

A deciduous perennial herb up to 12cm high, with a corm. The single basal leaf is linear and usually longer than the stem. Cream-coloured flowers are borne on a flattened, winged stalk; they have a red spot at the base of the smaller, lower segments and the undersides of the petals are pink; the flower tube is 1–3cm long.

Found in deep, red sand on the Coastal Plain and the Knersvlakte. **Flowers July to September.**

Lapeirousia barklyi

IRIDACEAE

A deciduous perennial herb up to 12cm
high, with a corm. The few linear to
sword-shaped leaves are ribbed; the
lowermost leaf is the longest, usually
exceeding the flowers. The flowers are
pale purple to pink with white or red
markings; the flower tube is long and
slender at the base, and widening and
curving higher up; the sharply pointed
petals are unequal in size, with the upper
petal larger and erect, the 3 lower ones
smaller and projecting horizontally.

Found in deep sand along the
northern Coastal Plain of Namaqualand
and in Namibia.
Flowers August to October.

PIETER VAN WYK

Lapeirousia exilis

IRIDACEAE

A deciduous perennial herb up to 5cm
high, with a corm. The linear, ribbed
leaves are crowded at ground level;
the longest leaf can reach 10cm. The
outer bracts on the flowering stalk are
channelled, erect, keeled and leaf-like.
Pale blue to white flowers are borne
close to ground level; they are marked
with darker blue at the midline or base
of the 3 slightly smaller, lower, horizontal
segments, and the uppermost segments
are erect.

Found on rocky hills in the
Namakwaland Klipkoppe and on the
Bokkeveld Escarpment.
Flowers June to August.

VERONICA ESTERHUIZEN

Lapeirousia fabricii
IRIDACEAE

A deciduous perennial herb up to 25cm high, with a corm. The basal leaf is ribbed and linear to sword-shaped. The bracts, borne on the flattened, branched flowering stalk, are much shorter. The few flowers are cream-coloured to yellow, marked with pale red on the 3 smaller, lower petals, while the undersides are flushed pink; the flower tube is slender, 3–5cm long.

Found on stone slopes throughout Namaqualand and also southwards to the Bokkeveld Mountains and Malmesbury in the Western Cape.
Flowers September to October.

ZELDA WAHL

Lapeirousia jacquinii
IRIDACEAE

A deciduous perennial herb up to 15cm high, with a corm. The few leaves are ribbed and linear to sword-shaped, the basal ones the longest. The green bracts on the flowering stalk are usually keeled, the margin toothed or crisped. The flowers are violet, the smaller, lower, horizontal petals with whitish to yellow markings and the upper petals spreading and erect; the flower tube is slender and long.

Found in sandy soil in the Namakwaland Klipkoppe and on the Knersvlakte, southwards to Worcester in the Western Cape.
Flowers August to September.

ZELDA WAHL

Lapeirousia macrospatha

IRIDACEAE   endemic

A deciduous perennial herb up to 10cm
high, with a corm. The 3–6 linear to sickle-
shaped leaves are ribbed, the lowermost
one usually much longer than the others.
The green bracts on the flowering stalk
are lightly keeled and pointed. The
flowers are cream-coloured to pink, the
slightly smaller, lower petals marked
with a small white to yellow basal patch
surrounded by a dark purple zone.

Found in deep sand on the Coastal
Plain of Namaqualand, from the Orange
River to Kleinzee.
Flowers August to September.

PIETER VAN WYK

Lapeirousia plicata

IRIDACEAE

A deciduous perennial, tufted herb up to
5cm high, with a corm. The several linear,
ribbed, lowermost leaves are about 1.5cm
long, the other leaves shorter. Whitish or
pale blue flowers, petals equal in size, are
borne at ground level; the petals often
have a dark blue to purple spot.

Found in loamy soil, sometimes with
small rocks, on flats in Namaqualand and
the rest of the Karoo.
Flowers July to August.

VERONICA ESTERHUIZEN

Lapeirousia silenoides
meidestert

IRIDACEAE   endemic

A deciduous perennial herb up to 15cm
high, with a corm. The usually single basal
leaf is linear to sword-shaped, up to 10cm
long and typically flat on the ground.
Bracts on the branched flowering stalk are
short, broad, keeled and have upcurved
tips. The numerous flowers are magenta
to cerise with creamy-yellow markings;
the flower tube is slender, curved and
3–5cm long.

   Found in coarse, granite-derived sand
and occasionally in crevices of granite
outcrops in the Namakwaland Klipkoppe.
Flowers July to September.

ANNELISE LE ROUX

Lapeirousia spinosa

IRIDACEAE   endemic

A deciduous perennial herb up to 10cm
high, with a corm. The usually single
basal leaf is linear, sharply pointed,
ribbed and 10–35cm long. The white or
cream-coloured flowers are sometimes
flushed pink, the midline of the smaller,
lowermost petals often marked pink to
purple and curving downwards.

   Found in red sand on flats in
Namaqualand, from the Richtersveld
southwards to Wallekraal.
Flowers August to October.

ZELDA WAHL

Lapeirousia tenuis

IRIDACEAE   endemic

A deciduous perennial herb up to 20cm
high, with a corm. The few linear to
sword-shaped leaves are ribbed, the
lowermost one the longest. Bracts on the
flowering stalk are folded or keeled near
the tips. The flowers are fragrant, pale
lilac to purplish, with petals narrowly
sword-shaped and pointed; the tube is up
to 1.5cm long.

Found on sandy flats, sometimes
among quartz pebbles, along the Coastal
Plain of Namaqualand.
Flowers August to September.

ZELDA WAHL

Lapeirousia verecunda

IRIDACEAE   endemic

A deciduous perennial herb up to 15cm
high, with a corm. The few linear leaves
are ribbed, the lowermost one the
longest. The flowering stalk is usually
branched, mostly at the base, and the
outer bracts are sword-shaped, keeled and
curved upwards at the tip. The flowers are
white, flushed pink on the outside, and
the lower petals have red, often heart-
shaped markings.

Found mostly on shale on the
escarpment west of Springbok.
Flowers August to September.

DENNIS HANSEN

ANNELISE LE ROUX

GLADIOLUS
There are about 170 species of *Gladiolus* in southern Africa of which 15 occur in Namaqualand.

Gladiolus arcuatus
IRIDACEAE   endemic

A deciduous perennial herb up to 35cm high, with a corm. The 3–5 basal leaves are linear and usually sickle-shaped, with hairs on the lower half, the raised midribs and margins. The 2–9 strongly fragrant, mauve or purple flowers are borne on a zigzag, arching stalk; the smaller, lower petals have a yellow to greenish lower half while the larger upper petal arches over the stamens.
    Found in sandy soil in Namaqualand and southwards to Clanwilliam in the Western Cape.
Flowers August to September.

PATRICK LANE

Gladiolus carinatus
blou-afrikaner
IRIDACEAE

A deciduous perennial herb up to 60cm high, with a corm. The base of the plant is purple with white mottling. The linear leaves have a prominent midrib; the margin is lightly thickened. The 4–8 flowers on the stalk are fragrant, blue to violet or yellow, occasionally pink; the smaller, lower petals often have transverse yellow markings and arch downwards; the upper middle petal is larger and arches over the stamens.
    Found in deep coastal sands in Namaqualand and southwards to Knysna in the Western Cape.
Flowers August to September.

Gladiolus caryophyllaceus
pienk-afrikaner

IRIDACEAE   endemic

A deciduous perennial herb up to 70cm
high, with a corm. The sword-shaped
leaves are hairy; the margin is thickened.
The large, strongly fragrant flowers are
pink to mauve and speckled or streaked
with darker colour.

   Found on sandy flats in Namaqualand,
southwards to Mamre, and in the
Swartberg.
Flowers August to October.

CAMERON MCMASTER

Gladiolus equitans
bergkalkoentjie, rooikalkoentjie

IRIDACEAE   endemic

A deciduous perennial herb up to 35cm
high, with a corm. The 3 or 4 basal leaves
are broadly sickle-shaped; the margin is
prominent and reddish. The 3–8 nodding,
orange to scarlet, sweetly scented flowers
are borne on a compressed stalk that
bends slightly downwards; the lower
petals bend downwards and have yellow
to greenish markings; the upper middle
petal is hooded over the stamens and the
upper lateral petals are spread sideways.

   Found in sandy and rocky areas in
the Namakwaland Klipkoppe and on the
Knersvlakte.
Flowers August to October.

VERONICA ESTERHUIZEN

JEAN SMITH

Gladiolus hyalinus
bruin-afrikaner

IRIDACEAE

A deciduous perennial herb up to 50cm high, with a corm. The 3 leaves are linear; the lowermost leaf is the longest and usually exceeds the flowering stalk; the margin and midrib are thickened. Yellowish to light brown flowers, often fragrant in the evenings, face in one direction; the lower petals are usually streaked and dotted with dark brown or purple; the basal half of the larger, upper middle petal is translucent and arched over the stamens.

Found on rocky slopes in the Namakwaland Klipkoppe and from the Bokkeveld Mountains to Port Alfred. **Flowers June to September.**

NINA HOBBHAHN

Gladiolus kamiesbergensis
Kamiesberg-kalkoentjie

IRIDACEAE   endemic

A deciduous perennial herb up to 50cm high, with a corm. The cylindrical leaves have four longitudinal grooves; the lowermost leaf is the longest and exceeds the flowering stalk. The flowers are mauve with violet speckles and flushed yellow on the lower petals; the upper middle petal arches over the stamens.

Found among rocks at higher elevations in the Kamiesberg only. **Flowers in October.**

Gladiolus orchidiflorus
groenkalkoentjie

IRIDACEAE

A deciduous perennial herb up to 45cm
high, with a corm. The 3–5 basal leaves
are linear to sword-shaped, curved or
straight, short and up to 6mm wide; the
margin and midrib are slightly thickened.
The 3–10 fragrant, hanging flowers are
dull green to pale purple, the lower
petals with dark purple and yellowish
markings. Petals are unequal: the upper
middle petal is the largest, 2–3.5cm long,
oblong to linear and is strongly arched
over the stamens; the 2 upper lateral
petals are spread sideways; the 3 lower
petals bend downwards.

Found in sandy, stony soil throughout
Namaqualand and also in parts of the
Western Cape, Free State and Namibia.
Flowers July to September.

ZELDA WAHL

Gladiolus saccatus
suursteen

IRIDACEAE

A deciduous perennial herb up to 70cm
high, with a corm. The narrow, linear,
sword-shaped leaves are prominently
3-ribbed, about 30cm long and grey-
green with white and purple mottling
near the base. The 4–10 flowers are
bright red and the flower tube is 2.5–3cm
long; the upper part of the flower tube is
cylindrical, with a conspicuous sac at the
base. The linear petals are unequal and
range from 6–10mm in length; the upper
middle petal is much longer, 35–40mm,
and extends horizontally.

Found on sandy or rocky flats and
slopes throughout Namaqualand,
extending southwards to Pakhuis and also
in Namibia.
Flowers May to August.

ANNELISE LE ROUX

Gladiolus salteri

IRIDACEAE   endemic

A deciduous perennial herb up to 30cm high, with a corm. The sword- to sickle-shaped leaves reach to approximately the top of the flowering stalk; the margin and midribs are thickened. The flowers are pale pink with deep pink and pale yellow markings on the lower petals; the upper middle petal is narrow and hooded over the stamens.

Found on rocky flats, sometimes in dorbank, east of Springbok in Namaqualand only.
Flowers August to September.

NOEL VAN ROOYEN

Gladiolus scullyi
kalkoentjie
IRIDACEAE

A deciduous perennial herb up to 45cm high, with a corm. The 4–7 basal leaves are linear to narrowly sword-shaped and 2–10mm wide. The 2–8 fragrant flowers are cream-coloured to pale yellow, the distal parts of the petals pale mauve, pink or purple. Petals are unequal: the upper middle petal is the largest and arched over the stamens, while the lower petals are united at the base; the flower tube is 1.2–1.7cm long.

Found on rocky slopes throughout Namaqualand and also in the drier parts of the Western Cape.
Flowers July to September.

ZELDA WAHL

Gladiolus viridiflorus
groenkalkoentjie
IRIDACEAE

A deciduous perennial herb up to 50cm high, with a corm. The 3 or 4 narrowly sword-shaped leaves are loosely twisted, with purple and white mottling near the base. The flowers are strongly fragrant and pale green; the lower petals have transverse yellow bands, outlined in purple, across the middle; the upper middle petal is the largest and is arched over the stamens.

Found on stony hills or mountain slopes in Namaqualand and southwards to Clanwilliam.
Flowers May to August.

ANNELISE LE ROUX

MELASPHAERULA
There is only one species of *Melasphaerula* in southern Africa.

Melasphaerula graminea
(formerly *Melasphaerula ramosa*)
feeklokkies
IRIDACEAE

A deciduous perennial herb up to 1m high, with a corm. The wiry stems usually trail through other bushes. The 3–6 leaves are linear, up to 40cm long and 2cm across. The numerous star-shaped, sour-smelling flowers are white to pale yellow and about 1.5cm in diameter; the lower petals are streaked with red-brown at the midline; the flowers are borne widely apart.

Found in moist places throughout Namaqualand, southwards to the Cape Peninsula and eastwards towards George.
Flowers June to October.

ZELDA WAHL & IVAN VAN NIEKERK

XENOSCAPA
There are three species of *Xenoscapa* in southern
Africa of which two occur in Namaqualand.

Xenoscapa fistulosa
IRIDACEAE

A deciduous perennial herb up to 15cm
high, with a corm. The 2 basal leaves are
soft-textured, oblong and lie flat on the
ground. The small, whitish to pale pink,
sweetly scented flowers have a narrow,
cylindrical, erect tube, which is 2–2.5cm
long; the upper middle petal is boat-
shaped and erect; the lower petals are
horizontal or bend downwards.

Found mostly in loamy soil, in the shade
of rocks or under shrubs, in the mountains
of Namaqualand and also in the Western
Cape, Eastern Cape and Namibia.
Flowers August to October.

ANNELISE LE ROUX

Xenoscapa uliginosa
IRIDACEAE   endemic

A deciduous perennial herb up to 5cm
high, with a corm. The 2 basal leaves are
soft-textured, oblong and lie flat on the
ground. The small, unscented, pale pink to
purple flowers have a narrow, cylindrical,
erect tube that is 2–3cm long, with
spreading petals; the lower 3 petals have
dark red and white markings.

Found in damp places in the shade of
rocks or under shrubs in the Kamiesberg.
Flowers August to October.

DAVE GWYNN-EVANS

SYRINGODEA
There are eight species of *Syringodea* in southern Africa of which one occurs in Namaqualand.

Syringodea longituba
IRIDACEAE

A deciduous perennial herb up to 10cm high, with a corm. The needle-like leaves, borne in a basal tuft, are often twisted and smooth or lightly hairy. The single flower is mauve, violet or violet-blue with a white to orange throat; the flower tube is 1.5–3cm long and very narrow.

Found in sandy soil in the Namakwaland Klipkoppe and also in the Bokkeveld Mountains and extending to Mossel Bay.
Flowers April to June.

RIAAN DE VILLIERS

GEISSORHIZA
There are about 95 species of *Geissorhiza* in southern Africa of which four occur in Namaqualand.

Geissorhiza kamiesmontana
IRIDACEAE   endemic

A deciduous perennial herb up to 30cm high, with a corm. The 3 or 4 nearly linear leaves reach to approximately the base of the flowering stalk; the margin is lightly raised. The flowers are blue to mauve with a white throat and a cylindrical tube 1.8–2.5cm long; the style reaches halfway up the anthers.

Found in the shade of rocks on the slopes of the Kamiesberg.
Flowers in September.

ANNELISE LE ROUX

IRIDACEAE **55**

ROMULEA
There are about 75 species of *Romulea* in southern Africa of which 17 occur in Namaqualand.

Romulea citrina
IRIDACEAE   endemic

A deciduous perennial herb up to 10cm high, with a corm. The 2 or more basal leaves curve outwards and are cylindrical, with narrow grooves. The green bracts have narrow, translucent margins; the inner bract margins are broader and streaked with brown. The flowers are lemon-yellow; the outer petals are often greenish or brownish on the back; the flower tube is short.

Found on slopes and flats, and often in seasonal pools, on the Kamiesberg. **Flowers August to September.**

Romulea kamisensis
IRIDACEAE   endemic

A deciduous perennial herb up to 10cm high, with a corm. The 2 or more basal leaves are long, cylindrical and narrowly 4-grooved. Bracts are green and the inner bracts have colourless, membranous margins. Magenta-purple flowers with violet stripes inside the throat are borne on a short stalk up to 5cm long; the white flower tube is narrowly funnel-shaped and longer than the petals, from 1.7–2.2cm in length; the stamens and style are included in the flower tube, with stigmas about as long as the anthers.

Found on rocky slopes, usually in shade, on the Kamiesberg. **Flowers July to September.**

Romulea minutiflora
IRIDACEAE

A small, deciduous perennial herb up to 10cm high, with a corm. The linear, basal leaves have narrow grooves and are often curved. The outer bracts have narrow, often brown-speckled, membranous margins; the inner bracts are almost completely membranous. The flowers are pale mauve or lilac, rarely whitish, often with a violet circle inside the throat; the cup is greenish-yellow and the outer petals are greenish or mottled on the back.

Usually found in disturbed areas of the Kamiesberg and also throughout the Western Cape.
Flowers July to September.

ANNELISE LE ROUX

Romulea namaquensis
IRIDACEAE   endemic

A deciduous perennial herb up to 20cm high, with a corm. There are usually 2 basal, cylindrical leaves with 4 narrow grooves. The inner bracts have wide, white or brown edges or brown-streaked, membranous margins and green, blunt tips. Shiny rose- to salmon-pink, rarely white, flowers have small, dark blotches or veins inside the throat; the filaments arise halfway up the flower tube and are maroon or sometimes yellow; the golden yellow anthers are about as long as the filaments.

Found on sandy or stony ground, sometimes in moist hollows, in the Namakwaland Klipkoppe.
Flowers July to September.

ZELDA WAHL

ZELDA WAHL

HESPERANTHA
There are about 80 species of *Hesperantha* in
southern Africa of which 11 occur in Namaqualand.

Hesperantha bachmannii
aandblom
IRIDACEAE

A deciduous perennial herb up to 30cm
high, with a corm. The 3 or 4 leaves are
linear to sword-shaped, soft, grass-like
and 10–15cm long, about half the length
of the slightly zigzag flowering stalk. The
few nodding, white flowers are about 2cm
in diameter; the petals bend backwards;
the flower tube is curved and about 10mm
long. The flowers open only in the late
afternoon and are then sweetly scented.

Found in damp, loamy soil, usually in
rocky shelters, in Namaqualand, extending
to the Western and Eastern Cape.
Flowers July to September.

TED OLIVER

Hesperantha flava
IRIDACEAE

A deciduous perennial herb up to 5cm
high, with a corm. The 2 or 3 leaves are
firm, oblong, sickle-shaped, up to 1cm
wide and blunt at the tip. The flowers are
trumpet-shaped, deep yellow inside and
brownish on the outside; the petals spread
horizontally. The flowers open at night
and are scented.

Found in stony, loamy soil in the
Namakwaland Klipkoppe, along the
inland escarpment to Matjiesfontein and
throughout the Karoo.
Flowers July to August.

Hesperantha flexuosa
IRIDACEAE   endemic

A deciduous perennial herb up to 20cm
high, with a corm. The 3 fleshy leaves are
linear to sickle-shaped and 0.5–1.5mm
wide. The 4–7 white flowers have a
straight, 5–10mm-long flower tube
and are borne on a zigzag stalk; the
horizontally spreading petals are longer
than the tube. The flowers open at night
and are scented.

Found in granite-derived sand, often in
wet sites, in the Namakwaland Klipkoppe.
Flowers August to September.

ZELDA WAHL

Hesperantha pauciflora
IRIDACEAE

A deciduous perennial herb up to 25cm
high, with a corm. The 3–5 soft, sword-
shaped leaves are 15–18cm long and
5–6mm wide. The flowers are pink to deep
mauve, with a straight tube 6–11mm long;
the petals are longer than the tube and are
borne on a zigzag stalk. The flowers open
in the afternoon and are usually unscented.

Found in seasonally wet, sandy soil
in the Richtersveld, the Namakwaland
Klipkoppe and also on the Bokkeveld
Plateau.
Flowers August to September.

ANNELISE LE ROUX

KOBUS KRITZINGER

BABIANA
There are about 90 species of *Babiana* in southern Africa of which 35 occur in Namaqualand.

Babiana carminea
IRIDACEAE   endemic

A stemless, deciduous perennial herb up to 10cm high. The pleated, sword-shaped leaves are velvety, hairy, crowded at the base and usually overtop the flowers. Bracts are covered with fine hairs; the inner bracts are forked at the tip. Deep mauve-red flowers with pale yellow nectar guides are borne at ground level; the flower tube is elongated and curved, the petals flaring open.

Found in rock crevices on limestone outcrops on the Knersvlakte.
Flowers July to August.

ANNELISE LE ROUX

Babiana curviscapa
perde-uintjie, bobbejaantjie
IRIDACEAE   endemic

A deciduous perennial herb up to 10cm high, with a corm. The 5–7 ribbed leaves are sword-shaped, often bent, up to 20cm long and 1.5cm across, and covered with short hairs. Bracts are covered by minute hairs; the inner bracts are forked at the tip. Numerous flowers are crowded on a stalk lying almost on the ground; they are magenta, purple or cherry-red, the lower lateral petals with spoon-shaped, white to yellow markings. The corms are edible.

Found in sandy, stony areas on slopes in the Namakwaland Klipkoppe.
Flowers July to September.

Babiana dregei

bobbejaantjie, klipuintjie

IRIDACEAE   endemic

A deciduous perennial herb up to 25cm
high, with a corm. The erect, rigid, sword-
shaped, hairless leaves are slightly pleated
and 1.2–2.5cm wide, with thickened,
straw-coloured veins; leaves with a sharp,
rigid tip. The compact stalk has numerous
flowers. Bracts are keeled, hairless or with
fine hairs, and are 3.5–6cm long, often
brown at the time of flowering; the inner
bracts are forked at the tip. Asymmetrical,
deep purple-blue flowers have long,
cylindrical tubes up to 6cm long, that
widen slightly below the throat; the 3
upper petals are slightly longer than the
lower ones. The corms are edible.

Found in rock crevices on hills and
mountain slopes in the Namakwaland
Klipkoppe.
Flowers August to October.

HEATHER BURGER

Babiana fimbriata

IRIDACEAE   endemic

A deciduous perennial herb up to 20cm
high, with a corm. Linear to lance-shaped
leaves, borne on a long, suberect leaf stalk,
are slightly pleated, loosely spirally twisted
or curled, sometimes with fine hairs on
the margins or veins. Bracts are green with
brownish margins and tips, hairless or finely
hairy on the midvein; the inner bracts are
split almost halfway. Flowers, borne on a
branched, hairless or sparsely hairy, zigzag
stalk, are mauve, the 2 lower lateral petals
marked with yellow and violet; the petals
are rounded at tips; the upper middle petal
is free, arched and widely separated from
the others; the 2 lower lateral petals have
slightly crisped margins.

Found on rocky slopes in the
Namakwaland Klipkoppe and on the
northern Knersvlakte.
Flowers July to September.

RUPERT KOOPMAN

IRIDACEAE **61**

HEATHER BURGER

Babiana flabellifolia
IRIDACEAE

A deciduous perennial herb up to 8cm high, with a corm. The wedge-shaped leaves are crowded at the base, irregularly toothed at the tip and appearing truncate; they can be hairless or finely hairy with velvety margins. Bracts are soft, somewhat membranous, hairless or minutely hairy; the inner bracts are split at the tips only. The flowers are widely bell-shaped and purple, the 2 lower lateral petals with pale to deep mauve and pale yellow patches; the flower tube is straight or slightly curved, cylindrical at the base and funnel-shaped towards the throat; the upper petal is slightly longer than the others.

Found in reddish soil on hills in the Namakwaland Klipkoppe, on the Knersvlakte and also on the Bokkeveld Plateau.
Flowers July to August.

PIETER VAN WYK

Babiana hirsuta
(formerly *Babiana thunbergii*)
sandlelie
IRIDACEAE

A deciduous perennial herb up to 60cm high, with a corm. The leaves are sword-shaped to linear, up to 30cm long, ribbed and minutely hairy to hairless. The inner bracts are forked at the tip. The numerous bright red flowers are 2-lipped, up to 5cm long and are closely set on 2–4 horizontal to ascending stalks; the upper petal stands erect and has inrolled margins, which envelop the filaments and style; the lower petals are partially fused to form a horizontal lip.

Found in white sand on the Coastal Plain and southwards to Saldanha.
Flowers August to October.

Babiana hypogaea

IRIDACEAE

A stemless, deciduous perennial herb up to 15cm high, with a corm. The lightly pleated, linear to sword-shaped leaves are usually erect, but can also bend backwards or lie on the ground. The leaves are sparsely to finely long-hairy, have firm tips and exceed the flowers. Sweetly scented, greenish-yellow to buff flowers, flushed brown to mauve on the outside, are borne at ground level; the flower tube is cylindrical, 3–4cm long and widens at the throat.

Found in red sand in the eastern Namakwaland Klipkoppe and also in Bushmanland.

Flowers June to September.

NOEL VAN ROOYEN

Babiana lewisiana

IRIDACEAE   endemic

A deciduous perennial herb up to 15cm high, with a corm. The broad, oblong, pointed, opposite leaves overlap, spread horizontally and clasp the erect stem; they are ribbed, only slightly pleated, with prominent veins, and are hairless or have minutely hairy veins and margins. The green bracts have dry, membranous tips and translucent margins; they are sparsely hairy and the inner bracts are split to about halfway. The flowers are magenta, the lower lateral petals greenish-yellow; all petals narrow towards the base and the upper middle petal is arched.

Found among quartz pebbles on the Knersvlakte.

Flowers June to August.

ANNELISE LE ROUX

PIETER VAN WYK

Babiana namaquensis
IRIDACEAE

A deciduous perennial herb up to 15cm high, with a corm. The linear leaves are spirally curled and twisted and may be slightly pleated; they are sparsely to very hairy and exceed the flowers. The hairless bracts have narrow, membranous margins and tips; the inner bracts are split at the tips and have fairly wide, membranous margins. Sweetly scented flowers are dark mauve to purple or almost white, the lower lateral petals with a yellow centre surrounded by white with dark mauve tips; the basal half of the flower tube is funnel-shaped and the outer petals end in stiff points.

Found in sandy soil on the northern Coastal Plain and also in Namibia. **Flowers June to August.**

RUPERT KOOPMAN

Babiana planifolia
IRIDACEAE   endemic

A deciduous perennial herb up to 30cm high, with a corm. The lance-shaped, stem-clasping leaves are sometimes loosely twisted distally; the wavy margins hairless or minutely hairy. The short bracts have a green basal half with translucent margins and are brown and dry towards the tips; the inner bracts are keeled and split to about halfway. Mauve flowers are borne on a horizontal stalk; the 2 lower lateral petals are pale yellowish-green below the mauve tip; the upper petal is free, widely separated from the others and arched at the base.

Found on rocky slopes throughout Namaqualand. **Flowers June to July.**

Babiana rubella

IRIDACEAE   endemic

A deciduous perennial herb up to 15cm high, with a corm. The stem-clasping, broadly sword-shaped leaves are held obliquely, or almost at right angles; they are slightly pleated, with dense, long hairs on the veins and margins. The green bracts are densely hairy with dry, rust-brown tips; the inner bracts are forked at the tip or to about halfway, and are transparent in the upper midline. Lightly rose-scented, pale to deep pinkish-mauve flowers are borne on an inclined stalk; the lower lateral petals are yellow with pink tips; the flower tube is about half the length of the petals; the lower petals are crisped along the bottom margins.

    Found on sandy flats on the Coastal Plain.

Flowers August to September.

RUPERT KOOPMAN

Babiana sinuata

IRIDACEAE

A deciduous perennial herb up to 35cm high, with a corm. The lance-shaped, twisted leaves have a fairly prominent midrib and wavy margins; they are sparsely hairy near the margins or only towards the base. The upper part of the bracts is keeled, hairless and membranous; the inner bracts are split halfway. Fragrant, pale mauve-blue flowers with yellow markings are spirally arranged on an erect or leaning stalk; the petals are blunt; the upper petal is free, bent backwards and widely separated from the others; the 3 lower petals are smaller and almost equal.

    Found on heavy, reddish, loamy soil on the Knersvlakte and southwards to Clanwilliam.

Flowers August to September.

RIAAN DE VILLERS

Babiana spiralis
IRIDACEAE

A deciduous perennial herb up to 55cm high, with a corm. The narrowly sword-shaped to linear leaves are lightly pleated, twisted above and hairless or lightly hairy on the margins or veins. The upper third of the green bracts is dry and tapered towards the tips; the inner bracts are slightly shorter and forked to about halfway. Blue-mauve to pink flowers with a strong freesia-like scent are borne on a velvety, branched stalk; the lower petals are sometimes yellow with blue-mauve or pink tips or completely yellow, as shown in the photograph; the flower tube is funnel-shaped and the upper middle petal arches initially, but later stands erect.

Found in deep sand on the Coastal Plain. **Flowers August to September.**

BRUCE WESSELS

Babiana striata
IRIDACEAE   endemic

A deciduous perennial herb up to 35cm high, with a corm. The upper half of the stem is curved slightly downwards. The lance-shaped, twisted leaves are recurved at the tip and slightly grooved; the margins deeply wavy. The short bracts have a green basal half with translucent margins and are brown and dry towards the tips; the inner bracts are keeled and split to about halfway. The flowers can be pale yellowish-green or pale mauve-purple, the lower lateral petals pale yellowish-green below the tip; the upper middle petal is widely separated from the others and arched at first.

Found on rocky hills throughout Namaqualand.
Flowers July to August.

RUPERT KOOPMAN

Babiana torta

IRIDACEAE   endemic

A deciduous perennial herb up to 8cm
high, with a corm. The very hairy, soft-
textured, lance-shaped, slightly wavy
leaves are arranged in a fan and spirally
curled towards the tips. The hairy bracts
have pale orange-brown tips; the inner
bracts are forked to almost halfway.
The fragrant flowers are pale mauve to
whitish, the lower lateral petals mainly
yellow with a purplish, arrow-shaped
outline on the lower half; the flower tube
is slightly curved and the upper middle
petal is larger and widely separated from
the other petals.
      Found on rock slabs in the Namakwa-
land Klipkoppe.
Flowers May to June.

IVAN VAN NIEKERK

TRITONIA
There are 28 species of *Tritonia* in southern Africa of
which four occur in Namaqualand.

Tritonia karooica

IRIDACEAE

A deciduous perennial herb up to 15cm
high, with a corm. The bent, sword-shaped
leaves are arranged in a fan. The heavily
scented flowers are yellow, flushed with
orange, and the petals are darkly veined;
the flower tube is longer than the petals;
the 3 lower petals have a thickened,
yellow median ridge.
      Found on dry, stony flats in the
Namakwaland Klipkoppe and southwards,
through the Roggeveld, to the Great Karoo.
Flowers August to September.

TED OLIVER

There are about 70 species of *Ixia* in southern Africa of which five occur in Namaqualand.

Ixia acaulis
IRIDACEAE   endemic

A stemless, deciduous perennial herb up to 3cm high, with a corm. The 3–5 basal, linear leaves are sometimes slightly sickle-shaped. Bracts are membranous; the outer bract is larger and somewhat fringed along the upper margin; the inner bract is similar in length, but narrower, 2-veined and lightly forked at the tip. The sweetly scented, bright yellow flowers have a whitish, elongated, cylindrical flower tube, with spreading petals.

Found in rock crevices on dolerite outcrops on the Knersvlakte.
Flowers May to June.

NICK HELME

Ixia namaquana
Namakwa kalossie
IRIDACEAE

A deciduous perennial herb up to 55cm high, with a corm. The leaves are sword-shaped, the uppermost leaf sheathing the stem. The outer bracts are pale, with 3 dark veins. The blue to pale mauve flowers are usually sweetly scented and half nodding; the yellowish flower tube is funnel-shaped; the petals range from the same length as the tube to twice as long.

Found on rocky, granite slopes in the mountains of the Richtersveld, the Namakwaland Klipkoppe and also in the Cederberg.
Flowers August to September.

ANNELISE LE ROUX

68   IRIDACEAE

Ixia ramulosa
kalossie
IRIDACEAE

A deciduous perennial herb up to 50cm
high, with a corm. The leaves are sword-
shaped. The outer bracts are pale, with 3
dark veins. The slightly nodding flowers
are purple, deep mauve or white; the
flower tube is narrowly funnel-shaped and
the petals are longer than the tube.
　　Found in sandy soil in the
Namakwaland Klipkoppe and also in the
Bokkeveld and Koebee Mountains.
Flowers September to November.

ANNELISE LE ROUX

Ixia scillaris
agretjie
IRIDACEAE

A deciduous perennial herb up to 50cm
high, with a corm. The 3–7 leaves are
sword-shaped; the lowest leaf is curved
and the uppermost one is suberect, up to
25cm long and usually strongly 3-veined.
The 7–25 scented flowers are borne on a
straight or slightly zigzag stalk with 1–3
short branches; they are pink to magenta,
occasionally white or mauve, and up to
2cm in diameter. The flower tube is very
slender and the petals are spreading;
the anthers hang down and are not
symmetrically arranged around the style.
　　Found on rocky slopes in the
Namakwaland Klipkoppe and from the
Bokkeveld Mountains southwards to
Somerset West in the Western Cape.
Flowers August to November.

ZELDA WAHL

IRIDACEAE **69**

ZELDA WAHL

Ferraria divaricata
krulletjie, spinnekopblom
IRIDACEAE

A deciduous perennial herb up to 40cm high, with a corm. The bluish-green leaves are entire, sheathed at the base and sword-shaped with sharp, rounded or incurved tips. Flowers about 5.5cm in diameter are borne on branched stalks; they are yellow to blue-green or brown, often with a purplish-blue or greenish median zone and brownish or greenish margins, which are wavy or crisped. The plant flowers for almost two weeks, but each flower remains open for only one day.

Found in sandy areas on the Coastal Plain, the Knersvlakte and also in the Western Cape.
Flowers August to November.

VERONICA ESTERHUIZEN

Ferraria ferrariola
krulletjie, spinnekopblom
IRIDACEAE

A deciduous perennial herb up to 60cm high, with a corm. The 3 or 4 lower leaves are long, slender and almost cylindrical, with narrow, often red-spotted sheaths; the shorter upper leaves are wider, sword-shaped and spreading. Sweetly scented flowers are borne on sparsely branched stalks, only partially covered with speckled leaf sheaths; they are pale greenish-yellow to greenish-blue, with pale greenish-grey margins; the 3 outer petals have fine, short, dark lines. The plant flowers for almost two weeks, but each flower remains open for only two to three days.

Found in stony, sandy or gravelly ground throughout Namaqualand and southwards to Clanwilliam.
Flowers June to September.

Ferraria foliosa
IRIDACEAE

A deciduous perennial herb up to 80cm high, with a corm. The basal leaves are linear and diamond-shaped in cross-section, with a thickened midline; the leaves on the stem spread horizontally and are channelled, without a central vein. The velvety, spotted, dark maroon or dark purple flowers have an unpleasant, mouldy odour; they are borne on a stalk with many branches arranged in a clockwise fashion. The plant flowers for almost two weeks, but each flower remains open for only one day.

Found in deep coastal sand, from just above the high-water mark to 15km inland, on the Coastal Plain and southwards to Elands Bay.
Flowers August to October.

ELIZE HOUGH

Ferraria ornata
IRIDACEAE   endemic

A deciduous perennial herb up to 8cm high, with a corm. The 4 or 5 grey-green, purple-tipped leaves are stem-clasping and have finely crisped, reddish, membranous margins; the basal leaves are relatively thick and finger-like; the upper leaves are broadly egg-shaped and concave. The leathery, grey-green bracts have translucent, membranous margins; the inner bracts are larger than the outer ones. The unscented flowers are white and pale yellow, speckled with brown, and have finely crisped, yellow and brown margins. The plant flowers for almost two weeks, but each flower remains open for only one day.

Found in deep, gritty, sandy soil on the Coastal Plain.
Flowers May to June.

GREGORY NICOLSON

VERONICA ESTERHUIZEN

Ferraria macrochlamys subsp. *kamiesbergensis*
krulletjie, spinnekopblom

IRIDACEAE   endemic

A deciduous perennial herb up to 18cm high, with a corm. The suberect, sword-shaped leaves have blunt ends and lack a midrib; the blades curve gently to one side; the margin is smooth and barely thickened. The unscented flowers are yellow with grey flecks or a whitish pale green with delicate green spots; the petal tips taper gradually towards long, frilly, coiled ends. The plant flowers for almost two weeks, but each flower remains open for only one to two days.

Found on rocky, granite slopes in the Namakwaland Klipkoppe.
Flowers August to November.

NINA SMITS

Ferraria macrochlamys subsp. *macrochlamys*
krulletjie, spinnekopblom

IRIDACEAE   endemic

A deciduous perennial herb up to 15cm high, with a corm. The blunt, sword-shaped leaves lack a midrib; the lower leaves are spread outwards and are often slightly sickle-shaped; the margin is thickened and crisped or wavy. The mildly putrid-smelling flowers are pale yellow, often with slightly dark yellow margins, the outer petals with bluish-green spots or a pale green median zone; the petal tips taper gradually towards long, frilly, coiled ends. The plant flowers for almost two weeks, but each flower remains open for only two to three days.

Found on rocky slopes in the Namakwaland Klipkoppe.
Flowers August to October.

Ferraria macrochlamys subsp. *serpentina*
krulletjie, spinnekopblom

IRIDACEAE   endemic

A deciduous perennial herb up to 15cm high, with a corm. The blunt, sword-shaped leaves lack a midrib; they are loosely folded back on themselves in concertina fashion; the margin is not thickened. The mildly putrid-smelling flowers are pale yellow, often with slightly dark yellow margins, the outer petals with bluish-green spots or a pale green median zone; the petal tips taper gradually towards long, frilly, coiled ends. The plant flowers for almost two weeks, but each flower remains open for only two to three days.

Found in loamy soil on the eastern side of the Coastal Plain of Namaqualand. **Flowers August to October.**

ANNELISE LE ROUX

Ferraria ovata
krulletjie, spinnekopblom

IRIDACEAE   endemic

A deciduous perennial herb up to 10cm high, with a corm. The basal leaves are linear, channelled below and up to 4.5cm long; the upper stem-clasping leaves are broadly egg-shaped with narrow, reddish, membranous margins. The 1 or 2 lightly acrid-scented, cream-coloured flowers have maroon streaks and a few brown blotches; the chocolate-brown margins are undulate and minutely crisped. The plant flowers for almost two weeks, but each flower remains open for only one day.

Found in stony soil in the Namakwaland Klipkoppe. **Flowers June to July.**

JOHN MANNING

NICOLA BERGH

Ferraria schaeferi
krulletjie, spinnekopblom
IRIDACEAE

A deciduous perennial herb up to 50cm high, with a corm. The overlapping, sword-shaped leaves arch outwards and are slightly fleshy. Yellow flowers with dark brown spots and margins are borne on a stalk, with branches arranged in a clockwise fashion; they are putrid to sweetly scented. The plant flowers for almost two weeks, but each flower remains open for only one day.

Found in deep coastal sand near the sea on the northern Coastal Plain and also on the Namibian coast.
Flowers August to September.

HESTER STEYN

Ferraria variabilis
krulletjie, spinnekopblom
IRIDACEAE

A deciduous perennial herb up to 20cm high, with a corm. The sword-shaped leaves are crowded at the base and lack a midrib; the margins often lightly thickened, rarely obscurely crisped; the leaves are usually as long as the flowering stalk, but can be up to twice as long. The often putrid-smelling flowers are dull yellow, yellow-green or light brown with banded or speckled markings and darker margins. The plant flowers for almost two weeks, but each flower remains open for only two to three days.

Found on sandy or shale flats and rocky slopes in Namaqualand, extending southwards to the Western Cape and also found in Namibia.
Flowers August to November.

There are about 175 species of *Moraea* in southern Africa of which 47 occur in Namaqualand.

Moraea ciliata
uintjie

IRIDACEAE

A deciduous perennial herb up to 20cm high, with a corm. The 3–5 greyish leaves are erect or curved and hairy; the margin fringed and often wavy. The spicy scented flowers are white, blue or yellow to light brown-pink with yellow markings; the 3 outer petals are large and have hairs in and around the markings; the 3 inner petals are smaller and narrower; the 3 style branches resemble petals. The flowers are short-lived.

Found in flat, sandy areas throughout Namaqualand and also in the Western Cape. **Flowers July to September.**

VERONICA ESTERHUIZEN, ALTA CHAMBERS & ANNELISE LE ROUX

Moraea deserticola
IRIDACEAE   endemic

A deciduous perennial herb up to 45cm high, with a corm. The 2 or 3 linear, channelled leaves are about as long as the erect flowering stalk, but are bent down to the ground; the margins become tightly inrolled when dry. The often nodding flowers are pale blue or whitish with yellow nectar guides on both the inner and outer petals; the filaments are completely united in a column; the style is concealed by the anthers.

Found on or at the foothills of dolomite hills on the Knersvlakte. **Flowers June to August.**

ANNELISE LE ROUX

Moraea falcifolia
IRIDACEAE

A deciduous, stemless perennial herb up to 5cm high, with a corm. The spreading, channelled, twisted leaves have undulate margins. The flowers are white with yellow nectar guides at the centre of the outer petals and purple markings on the narrower, inner petals; the 3 erect style branches in the centre of the flower resemble petals.

Found on sandy or open clay flats in Namaqualand, southwards to the Bokkeveld Mountains and extending to the Western Cape, also in Namibia. **Flowers May to August.**

ANNELISE LE ROUX

Moraea fenestralis
IRIDACEAE

A deciduous, stemless perennial herb up to 2.5cm high, with a corm. The 3 or 4 bent, cylindrical, fleshy leaves have a longitudinal band of transparent tissue on the upper surface. The flowers are pale pink to lilac with a yellow centre; the filaments are united in a column.

Found in shallow soil on granite outcrops in the Namakwaland Klipkoppe. **Flowers June to July.**

TED OLIVER

Moraea fugacissima
IRIDACEAE

A deciduous, stemless perennial herb up to 5cm high, with a corm. The linear to cylindrical, channelled leaves are erect to spreading. The yellow flowers are tubular at the base and the petals form a cup; the filaments are united in a column and the style exceeds the anthers.

Found in damp sand in the Namakwaland Klipkoppe and also in the mountains of the Western Cape, extending to Humansdorp in the Eastern Cape. **Flowers July to September.**

ANNELISE LE ROUX

Moraea filicaulis
(formerly *Moraea fugax* subsp. *filicaulis*)
klipuintjie
IRIDACEAE

A deciduous perennial herb up to 60cm high, with a corm. The 1 or 2 cylindrical leaves are keeled, trailing and much longer than the stem. The strongly scented flowers, about 4cm in diameter, are white, blue or yellow with yellow patches on the outer petals. Usually only one flower is open at a time.

Found in flat, sandy areas in Namaqualand and also in the Western Cape. **Flowers August to December.**

ZELDA WAHL

Moraea herrei
IRIDACEAE   endemic

A deciduous perennial herb up to 10cm high, with a corm. The single, cylindrical leaf is erect to spreading and loosely spiralled. The flowers are enclosed by firm, green bracts and are blue to purple with small, yellow patches on the throat; the subequal petals are spreading. An abundance of this species in an area usually indicates disturbance.

Found on rocky slopes and flats in the mountainous areas of Namaqualand. **Flowers September to October.**

HEATHER BURGER

Moraea inconspicua
taaiuintjie
IRIDACEAE

A deciduous perennial herb up to 45cm high, with a corm. The 1 or 2 flat, sometimes channelled leaves are usually longer than the stem and straight to slightly spiralled or tightly coiled. Small inconspicuous flowers are borne on a much-branched stalk that is sticky below the nodes; they are cream-coloured or dull yellow to dark brown with yellowish patches on the outer petals; the petals bend down. The flowers open just after noon.

Found in sandy, rocky places in the Namakwaland Klipkoppe and southwards to the Western and Eastern Cape. **Flowers September to November.**

ANNELISE LE ROUX

Moraea knersvlaktensis

IRIDACEAE   endemic

A deciduous perennial herb up to
25cm high, with a corm. The 5 or more
channelled leaves have undulate margins.
Pale yellow flowers, with darker yellow
patches outlined in green, are borne on a
much-branched, reddish stalk; the slightly
twisted petals spread horizontally and are
slightly wavy; the filaments are united in a
column about 6mm long and the anthers
are spread out.

   Found widely scattered among quartz
pebbles, on clay and stony ground, on the
Knersvlakte.
Flowers July to September.

Moraea lewisiae

IRIDACEAE

A deciduous perennial herb up to 90cm
high, with a corm. The 1–3 channelled
leaves have straight, inrolled margins
and are whip-like and trailing. Yellow
flowers are borne on a long stalk;
the style splits into 6 cylindrical arms
extending between the stamens. The
flowers open at about 18:00.

   Found on stony soil on flats or slopes
in the Namakwaland Klipkoppe and
southwards to the Western Cape.
Flowers October to November.

Moraea miniata
tulp
IRIDACEAE

A deciduous perennial herb up to 60cm high, with a corm. The 2 leaves are narrow, linear and usually longer than the flowering stem. Yellow to salmon-pink, occasionally white flowers, up to 4.5cm in diameter and with minutely speckled yellow patches, are borne on a branched stalk; the filaments are united in a column with a swollen base and the style is concealed by the anthers.

Found in flat, sandy areas throughout Namaqualand, southwards to the Western Cape and also in the Calvinia area. **Flowers September to October.**

ZELDA WAHL

Moraea nana
IRIDACEAE

A deciduous perennial herb up to 30cm high, with a corm. The 3 or 4 narrowly channelled leaves can be up to 40cm long and occur well above the ground. Pale yellow to brownish-salmon flowers are borne on a stalk that starts branching just above the leaves; the inner petals are slightly smaller than the outer ones; the style splits into 6 cylindrical arms extending between the stamens.

Usually found in deep, sandy soil in the Namakwaland Klipkoppe and on the Knersvlakte, southwards to Citrusdal. **Flowers September to October.**

IVAN VAN NIEKERK

Moraea pendula

IRIDACEAE   endemic

A deciduous perennial herb up to 75cm
high, with a corm. The 3 or 4 channelled
leaves are longer than the stem. Hanging
flowers are borne on a much-branched
stalk; they are pale yellow, occasionally
pink with a minutely speckled, slightly
darker yellow centre; the petals bend
completely backwards; the filaments are
united in a column with a swollen base;
the style is concealed by the bright orange
to red anthers.

Found only in seasonally poorly
drained soils on the Kamiesberg plateau.
Flowers August to October.

ANNELISE LE ROUX

Moraea pilifolia

IRIDACEAE

A stemless, deciduous perennial herb up
to 5cm high, with a corm. Often found
in small clumps. The channelled, blunt,
oblong leaves bend slightly backwards;
the margins are thickened, membranous
and covered with dense, long, prickly hairs.
Yellow flowers are arranged in a basal tuft;
the filaments are united and the anthers
almost reach the fringed stigmas.

Found only in loamy, stony soil in
the Namakwaland Klipkoppe, on the
Knersvlakte and also in the Giftberg.
Flowers June to July.

IVAN VAN NIEKERK

Moraea schlechteri
tulp

IRIDACEAE   endemic

A deciduous perennial herb up to 30cm high, with a corm. The blue-green, channelled, sword-shaped leaves have a broad, sheathing base and can be as long as the stems. Yellow flowers, up to 3.5cm in diameter and usually with darker yellow patches, are borne on a widely branched stalk; the filaments are united in a straight column 4–5mm long, but are free and spreading at the tips; the style is longer and branched at the end.

Found in flat, sandy, stony places in the Namakwaland Klipkoppe.
Flowers July to September.

Moraea serpentina
slanguintjie

IRIDACEAE

A deciduous perennial herb up to 15cm high, with a corm. There are usually 3 long, very narrow and spirally twisted basal leaves. The few white to yellow flowers, about 3.5cm in diameter and sometimes flushed mauve or pink, are borne on a sparsely branched stalk; inner petals are smaller and stand erect.

Found in sandy, stony places throughout Namaqualand and also in Bushmanland and the Olifants River Valley.
Flowers September to October.

Moraea setifolia
papieruintjie
IRIDACEAE

A deciduous perennial herb up to 20cm high, with a corm. The 1 or 2 long, thin leaves are curved and channelled. The flowers are pale blue-grey with yellow to orange and white patches on the outer petals; the petals bend backwards; the filaments are united at the base.

Found in coarse sand, often in disturbed sites, in the Namakwaland Klipkoppe and on the Knersvlakte, extending southwards and eastwards to Grahamstown.
Flowers September to October.

ANNELISE LE ROUX

Moraea tortilis
poepuintjie

IRIDACEAE   endemic

A deciduous perennial herb up to 15cm high, with a corm. The 2 or 3 long, basal leaves are narrow, cylindrical and coiled like a corkscrew. The blue to white flowers have white patches with small, yellow markings on the outer petals; the petals bend backwards.

Found in sandy or stony areas in Namaqualand.
Flowers September to October.

ZELDA WAHL

IRIDACEAE **83**

BULBINE
There are about 45 species of *Bulbine* in southern Africa of which 32 occur in Namaqualand.

Bulbine abyssinica
ASPHODELACEAE

A deciduous perennial herb up to 60cm high, forming large, compact tufts. The fleshy, cylindrical leaves have yellowish, papery basal sheaths. Many yellow flowers with fluffy stamens are borne on an erect stalk, the youngest flowers and buds at the tip.

Found on flats and slopes on the Coastal Plain and also throughout South Africa, into tropical Africa.
Flowers August to November.

Bulbine alooides
kopiva, wildekopiva
ASPHODELACEAE

A deciduous perennial herb up to 40cm high, with swollen roots. The 6–12 fleshy, basal leaves are up to 20cm long and 2.5cm across and taper towards the tip; the margin has short hairs; they emerge only at flowering. Many yellow flowers with fluffy stamens are borne on an erect stalk, the youngest flowers and buds at the tip.

Found on rocky flats from Namaqualand southwards to Worcester and also on the Bokkeveld and Roggeveld Escarpments.
Flowers March to May.

Bulbine brunsvigiaefolia

ASPHODELACEAE   endemic

A perennial herb up to 70cm high, with a
tuber. The 5–8 broad, pale green, fleshy
leaves are faintly veined; the margin is
densely fringed. Many yellow flowers with
fluffy stamens are borne on an erect stalk,
the youngest flowers and buds at the tip.
    Found in sand or clay among rocks in
the Namakwaland Klipkoppe.
Flowers in August.

Bulbine dactylopsoides

ASPHODELACEAE   endemic

A deciduous perennial herb up to 30cm
high, proliferating from secondary tubers.
The 2–4 finger-like, fleshy, greyish-green
or reddish leaves have basal sheaths; they
are borne on a small stem, up to 2cm
above ground when flowering. The few
yellow flowers with fluffy stamens are
borne on an erect stalk, the youngest
flowers and buds at the tip.
    Found among quartz pebbles on
the Knersvlakte.
Flowers May to August.

Bulbine diphylla
ASPHODELACEAE

A deciduous perennial herb up to 15cm high, proliferating from secondary tubers. The 2 or 3 fleshy, ovoid, greyish-green or reddish-brown leaves are borne on a small stem, up to 2cm above ground when flowering. The few yellow flowers, with fluffy stamens and with petals bent backwards, are borne on an erect stalk, the youngest flowers and buds at the tip.

Found among quartz pebbles on the Knersvlakte and also in the Pakhuis Mountains.
Flowers May to August.

ANNELISE LE ROUX

Bulbine fallax
ASPHODELACEAE   endemic

A deciduous perennial herb up to 30cm high, with a tuber. The 12–20 fleshy leaves are pressed flat on the ground; upper leaf surface with bold patterns of several broad, rectangular windows bordered in pale green; the margin is minutely hairy; the leaves wither and turn orange at flowering. The yellow flowers, with fluffy stamens and with petals bent backwards, are borne on an erect stalk, the youngest flowers and buds at the tip.

Found in hard, loamy soil in the Namakwaland Klipkoppe and on the Knersvlakte.
Flowers August to September.

ANNELISE LE ROUX

Bulbine frutescens
kopiva, wildekopiva
ASPHODELACEAE

A shrub up to 50cm high, with fibrous roots. The fleshy, cylindrical leaves are bright green, about 4–8mm thick, 15cm long and surrounded by hard, grey sheaths at the base. Many yellow flowers with fluffy stamens are borne on an erect stalk, the youngest flowers and buds at the tip.

Found throughout Namaqualand and other parts of the Karoo.
Flowers September to April.

ZELDA WAHL

Bulbine louwii
ASPHODELACEAE

A perennial herb up to 15cm high, with a solitary, round tuber. The 25–30 leaves are arranged in a tight, spreading rosette on the ground; the upper surface is somewhat concave and slightly transparent; the margin is coarsely fringed. Many pale yellow flowers with fluffy stamens are borne on a short stalk, the youngest flowers and buds at the tip.

Found in open patches among quartz pebbles on the Knersvlakte and the western edge of Bushmanland.
Flowers July to August.

ANNELISE LE ROUX

ASPHODELACEAE **87**

Bulbine margarethae

ASPHODELACEAE   endemic

A perennial herb up to 15cm high, with many round tubers. The leaves, arranged in a rosette, are bright green or greyish to reddish-green with whitish lateral and longitudinal veins connected together like the mesh of a net; the margin and midvein are smooth or occasionally fringed with hairs. Pale yellow flowers, with fluffy stamens and with petals bent backwards, are borne on an erect stalk, the youngest flowers and buds at the tip.

Found in dolomite rock crevices on the Knersvlakte.
Flowers August to September.

ANNELISE LE ROUX

Bulbine mesembryanthoides
waterkannetjie
ASPHODELACEAE

A perennial herb, sunken below ground level, with a small tuber. The 2 or 3 leaves have a flat, translucent tip at ground level. Pale yellow flowers, with 3–6 widely spaced fluffy stamens and with petals bent backwards, are borne on an erect stalk up to 10cm long.

Usually found in patches among small, white quartz pebbles in Namaqualand and the Little Karoo.
Flowers August to November.

HERGEN EHRHARDT

Bulbine ophiophylla
ASPHODELACEAE

A perennial herb up to 15cm high, with rhizomes and orange roots. The old, fibrous leaf bases form an oblong to elliptic false stem 3–4cm high. The brittle, twisted, flattened, fleshy leaves are grey-green to dark olive-green, with a rounded maroon border. Slightly scented, pale peach to pale yellow flowers, with fluffy stamens and with petals curved backwards, are borne on a stalk up to 10cm long. The flowers open successively 1 or 2 at a time, early in the morning.

Found in sandy areas on the northern part of the Coastal Plain, northwards to Namibia.
Flowers May to August.

ZELDA WAHL

Bulbine praemorsa
blougif
ASPHODELACEAE

A deciduous perennial herb up to 60cm high, with a tuber. The 2–10 fleshy, sheathing leaves are deeply channelled above and about 30cm long. Many yellow flowers with fluffy stamens are borne on an erect stalk, the youngest flowers and buds at the tip.

Found in red, sandy, loamy soil throughout Namaqualand, southwards to Worcester and the Little Karoo.
Flowers July to September.

ANNELISE LE ROUX

ASPHODELACEAE **89**

Bulbine quartzicola
ASPHODELACEAE   endemic

A perennial herb up to 18cm high, with
a tuber. The 2–6 erect, cylindrical, fleshy,
3–5cm-long leaves are tightly clasped at
the base to forms a smooth, solid neck of
white to dark, papery sheaths; the leaf
withers and forms a tight spiral or curl at
the tip during flowering. Yellow flowers
with fluffy stamens are borne on an erect
stalk, the youngest flowers and buds at
the tip.

Found among white quartz pebbles in
Namaqualand.
Flowers September to November.

Bulbine sedifolia
ASPHODELACEAE   endemic

A deciduous perennial herb up to 30cm
high, with a small tuber. The 3–10 fleshy,
deeply channelled leaves are pale green
with a grey, metallic sheen; they arise
in a tuft from a short, bristly neck or
membranous sheath that is up to 2cm
long; the tip withers quickly. Yellow
flowers with fluffy stamens are borne on a
stalk that bends outward at the base and
then becomes erect.

Found in stony soil in the Namakwaland Klipkoppe.
Flowers July to September.

Bulbine torta
kopiva, wildekopiva
ASPHODELACEAE

A deciduous perennial herb up to 20cm high, with a tuber. The 4–20 basal, twisted, thread-like leaves are at ground level. Yellow flowers with fluffy stamens are borne on an erect stalk, the youngest flowers and buds at the tip.

Found in sandy or rocky, dry areas in the Namakwaland Klipkoppe, southwards to the Bokkeveld, Cederberg, Hantam and Klein Roggeveld Mountains. **Flowers July to September.**

RIAAN DE VILLIERS

Bulbine vittatifolia
ASPHODELACEAE   endemic

A deciduous perennial herb up to 8cm high, with a tuber. The 15–20 fleshy, slightly sickle-shaped leaves are subcylindrical; the upper surface is flattened and has longitudinal white lines. Sweetly scented yellow flowers with fluffy stamens are borne on an erect, slightly leaning stalk, the youngest flowers and buds at the tip.

Found between granite boulders on sandy plains in the Namakwaland Klipkoppe. **Flowers March to May.**

ANNELISE LE ROUX

ASPHODELACEAE **91**

ZELDA WAHL

BULBINELLA
There are about 20 species of *Bulbinella* in southern Africa of which nine occur in Namaqualand.

Bulbinella cauda-felis
witkatstert
ASPHODELACEAE

An erect perennial herb up to 80cm high, with a corm. The basal leaves are greyish-green, erect, channelled, up to 60cm long and 1cm wide; the base of the outer leaves dilates to form a membranous, cream-coloured, sometimes reddish sheath. The stalk, up to 20cm long, bears 50–150 tightly packed pure white flowers.

Found in loamy soil in the Namakwaland Klipkoppe, on the Knersvlakte and throughout the Western Cape.
Flowers August to December.

Bulbinella ciliolata, endemic to the Namakwaland Klipkoppe, also has white flowers and is similar in appearance.

RIAAN DE VILLIERS

Bulbinella divaginata
katstert, geelpiet
ASPHODELACEAE

An erect, deciduous perennial herb up to 45cm high, with rhizomes and loose to compact fibres (old leaf bases) up to 8cm long covering the new leaf bases. The 4–10 thin, thread-like leaves can reach a length of 50cm; they appear at the start of flowering, but will develop fully only some three weeks later. Bright yellow flowers are borne on a thin stalk.

Found in seasonally moist, sandy or loamy soil on the Coastal Plain and in the Namakwaland Klipkoppe, southwards to the Western Cape.
Flowers March to June.

Bulbinella gracilis

ASPHODELACEAE   endemic

A deciduous perennial herb up to
30cm high, with rhizomes and 1 or 2
membranous sheaths covering the new
leaf bases. The many fleshy, cylindrical
leaves are hairless. Small, yellow flowers
are borne close together at the end of
stalks longer than the leaves.

Found in cool, moist places, usually
under large rocks, on south-facing slopes
throughout Namaqualand.
Flowers June to August.

PIETER VAN WYK

Bulbinella latifolia

geelkopmannetjie, heuningblom

ASPHODELACEAE

An erect perennial herb up to 1m high,
with rhizomes and fine to medium thick
loose fibres (old leaf bases) covering the
new leaf bases. The 5–10 basal leaves
are bright green, erect to suberect,
channelled, hairless and slightly fleshy;
they vary in size and can become up to
50cm long and 6.5cm wide; the margin
is entire or minutely toothed. About 500
tightly packed, bright yellow flowers are
borne at the distal end of an erect stalk.
Poisonous to stock.

Found in seasonally damp places in the
Namakwaland Klipkoppe, southwards to
the Cederberg.
Flowers August to October.

ANNELISE LE ROUX

ANNELISE LE ROUX

TRACHYANDRA
There are about 50 species of *Trachyandra* in southern Africa of which 20 occur in Namaqualand.

Trachyandra bulbinifolia
solknol
ASPHODELACEAE

A perennial herb up to 40cm high, with rhizomes. The numerous basal leaves are greyish-green, cylindrical, fleshy and up to 18cm long and 5mm across; they are smooth or fringed and often curled when young. A collar of pale, fringed scales surrounds the plant at the base. White flowers, about 1.2cm in diameter and spotted at the base, are borne on a loosely branched stalk.

Found on sandy to rocky flats throughout Namaqualand, extending southwards to Saldanha Bay and northwards into Namibia.
Flowers July to October.

ZELDA WAHL

Trachyandra falcata
veldkool, Khoi-kool
ASPHODELACEAE

A perennial herb up to 60cm high, with rhizomes. The 4 or 5 basal leaves are flat, curved and leathery, up to 30cm long and 3–5cm across; they are minutely hairy or hairless. Pale mauve to white flowers, about 1cm long and spotted with yellow at the base, are borne densely packed on a branched stalk. The very young unopened flowers are used in stews.

Found in sandy soil in stony areas throughout Namaqualand, extending southwards to Saldanha Bay and northwards into Namibia.
Flowers July to October.

94 ASPHODELACEAE

Trachyandra flexifolia
ASPHODELACEAE

A perennial herb up to 25cm high, with rhizomes. The 2–6 basal leaves are straight or wavy, about 1mm across, and usually very hairy. White, dark-keeled flowers, with yellow spots near the base of the petals, are borne on a hairy, branched stalk.

Found on sandy and shale flats and slopes in the Namakwaland Klipkoppe, on the Knersvlakte and also in the Western Cape.
Flowers May to September.

ANNELISE LE ROUX

Trachyandra karrooica
ASPHODELACEAE

A perennial herb up to 5cm high, with rhizomes. The fleshy, cylindrical leaves are about 1mm across and round in cross-section; they are hairy or hairless. Scented flowers are borne on widely spreading, branched stalks; the white petals have a brown median stripe and yellow spots near the base. The flowers open in the afternoon.

Found in stony soil in Namaqualand and also in the Little Karoo, Great Karoo and Namibia.
Flowers February to April.

ANNELISE LE ROUX

ASPHODELACEAE **95**

ZELDA WAHL

Trachyandra muricata
beesblom
ASPHODELACEAE

A perennial herb up to 50cm high, with rhizomes. The 3–10 basal, flat leaves are up to 80cm long and 0.5–5cm across; the margin is rough and minutely spiny. White flowers, up to 2cm in diameter and with paired yellow spots at the base of the petals, are borne on a widely branched stalk. Cattle are fond of eating this plant.

Found on hills throughout Namaqualand and also in the Western Cape and Namibia.
Flowers July to October.

ANNELISE LE ROUX

Trachyandra paniculata
ASPHODELACEAE

A perennial herb up to 40cm high, with rhizomes. The 3 or 4 slightly fleshy, flat leaves are up to 1cm across, straight or wavy and tapered towards the ends; the margin is raised, usually minutely fringed. White flowers, spotted with yellow near the base of the spreading petals, are borne on a branched stalk.

Found on stony slopes in the Richtersveld and on the Knersvlakte, extending southwards to Wellington, also in Namibia.
Flowers August to October.

Trachyandra revoluta
ASPHODELACEAE

A perennial herb up to 80cm high, with rhizomes. The erect young and prostrate older leaves are cylindrical, slightly fleshy and up to 40cm long, sometimes slightly undulate or with a spiral twist. Hanging flowers are borne on a slender, much-branched stalk; the white petals have a brown stripe, are spotted with yellow near the base and bend completely backwards.

Found on sandy or clay flats throughout Namaqualand, southwards to the Cape Peninsula and eastwards to East London. **Flowers July to October.**

Trachyandra tortilis
ASPHODELACEAE

A perennial herb up to 25cm high, with rhizomes. The 3–6 greyish-green, basal leaves are folded transversely in concertina fashion; they are slightly fleshy, flat to round, and up to 10cm long. Pale pink flowers, up to 2cm in diameter and marked with green, are borne on a much-branched stalk. The flowers open only in the afternoon.

Found on sandy and clay flats in Namaqualand and southwards to Hopefield in the Western Cape. **Flowers June to August.**

There are five species of *Aloidendron* in southern Africa of which three occur in Namaqualand.

Aloidendron dichotomum
(formerly *Aloe dichotoma* var. *dichotoma*)
quiver tree, kokerboom
ASPHODELACEAE

A succulent tree up to 9m tall, branched from about halfway up. Young, smooth branches fork repeatedly, forming densely rounded crowns; the bark forms large, golden brown scales with sharp edges. The blue-green, narrow, oblong, fleshy leaves are up to 30cm long; the margin has short, broad, yellowish teeth 1mm long. Yellow flowers are borne close together on an erect, branched stalk. San people once used the hollowed stems to carry their arrows.

Usually found on north-facing slopes in Namaqualand, eastwards to the Upington area and northwards into Namibia. **Flowers June to August.**

Aloidendron pillansii
(formerly *Aloe pillansii*)
basterkokerboom
ASPHODELACEAE

A large, succulent tree up to 12m tall. The few robust, erect branches have large rosettes of leaves at the ends; the bark is usually smooth and whitish, but is somewhat coarse in very old specimens. The greyish-green, fleshy leaves distinctly clasp the branches and are fairly large, up to 60cm long and 10cm wide at the base; the margin is armed with small, white teeth. Yellow flowers are borne on a branched stalk that recurves downward and arises from the lowest leaves of the rosettes.

Found on stony hillsides in the Richtersveld, extending into southern Namibia.
Flowers in October.

Aloidendron ramosissimum
(formerly *Aloe dichotoma* var. *ramosissima*)
bush quiver tree, nooienskokerboom
ASPHODELACEAE

A shrubby succulent up to 3m tall. Probably the most profusely branched of all aloes; the smooth stems end in relatively small rosettes of narrow, oblong leaves. The fleshy leaves are up to 20cm long and 2cm wide at the base; the margin has very small, brownish teeth. Yellow flowers are borne on an erect, branched stalk up to 20cm long.

Found on rocky hills and mountains in the Richtersveld and also in Namibia. **Flowers June to August.**

Very similar to the quiver tree, *Aloidendron dichotomum*, but generally recognised by the absence of a main trunk.

ANNELISE LE ROUX

ALOE
There are about 130 species of *Aloe* in southern Africa of which 18 occur in Namaqualand.

Aloe arenicola
bont-o-t'korrie
ASPHODELACEAE

A branched or unbranched perennial succulent up to 40cm high. The flat, sword-shaped, fleshy, blue-green leaves have scattered, oblong white spots, are up to 18cm long and 5.5cm wide and sheathe the stem at the base; the margin is white and horny, with minutely spaced teeth. Reddish, hanging flowers up to 4cm long are borne at the tip of a branched stalk.

Found on the sandy coastal belt within a few hundred metres from the beach, from the Orange River to Lambert's Bay. **Flowers July to September.**

ZELDA WAHL

Aloe falcata
ASPHODELACEAE

A succulent up to 50cm high. Forms dense clumps with rosettes pointing outwards, lying almost on their side, with leaves curving upwards. The green to greyish-green, fleshy leaves are firm in texture with rough, sandpaper-like surfaces; the margin has pale to dark brown teeth. Tubular, hanging flowers, usually dull red, rarely yellow, are borne on an erect, branched stalk; the stamens are longer than the petals.

Found in dry, flat, sandy areas or in shallow soil on rocky outcrops in Namaqualand and eastwards to Calvinia. **Flowers in December.**

ANNELISE LE ROUX

Aloe framesii
(formerly *Aloe microstigma* subsp. *framesii*)
bitter aloe, bitteraalwyn
ASPHODELACEAE

A succulent up to 1m high. Forms dense colonies, 2–3m in diameter, with branched, horizontal and inconspicuous stems. The fleshy, narrow, sword-shaped leaves are up to 30cm long and 7cm across, dull grey, usually with white spots; the margin has triangular teeth, about 3mm long. Tubular orange-red flowers with a greenish-yellow tip are borne in an oblong, pointed arrangement at the ends of an erect stalk up to 80cm high and branched 2 or 3 times; they hang down with age.

Found in the sandy coastal belt, from Port Nolloth southwards to near Saldanha Bay.
Flowers June to July.

ZELDA WAHL

Aloe gariepensis
Orange River aloe, Oranjerivier-aalwyn
ASPHODELACEAE

A succulent up to 1m high. Usually solitary, varying in size from small and stemless to a larger plant with a short stem. Leaves with a characteristic striped appearance are copiously spotted on both surfaces in young plants; some spots remain on the upper surface as the plant matures; the brown, horn-like margin has small, sharp, triangular teeth. Yellow to greenish-yellow flowers, or sometimes bicoloured with red buds and yellow when open, are closely packed on a stalk up to 1m high; they are tubular and slightly wider towards the mouth; young buds are more or less hidden by long bracts.

Found on steep, rocky slopes or in rock crevices close to the Orange River in the Richtersveld, extending into southern Namibia.
Flowers July to September.

ZELDA WAHL & PIETER VAN WYK

Aloe glauca
blouaalwyn
ASPHODELACEAE

A stemless or short-stemmed succulent up to 60cm high. The distinctly blue-grey, fleshy leaves, up to 40cm long, have faint longitudinal lines; the lower leaf surface often has small, scattered spines towards the tips; the margin has contrasting reddish-brown teeth. Pink to pale orange flowers are borne on a stout unbranched stalk; they are upright in the bud stage, while the mature open flowers hang down.

Found mostly on rocky hills and mountain slopes in the Namakwaland Klipkoppe and southwards to Laingsburg and Swellendam.
Flowers August to October.

NOEL VAN ROOYEN

Aloe khamiesensis
(formerly *Aloe microstigma* subsp. *microstigma*)
Kamiesberg aloe, Kamiesberg aalwyn
ASPHODELACEAE

A small succulent, but can be up to 2m high. Usually single-stemmed with a rosette, but sometimes with 2 branches. The rich green, fleshy, sword-shaped, tapering leaves have roundish white marks and are up to 40cm long and 8cm wide at the base; the margin has reddish-brown triangular teeth. Tubular, orange-red flowers with greenish-yellow tips are crowded at the top of a 4–8-branched stalk.

Found on rocky outcrops in the Namakwaland Klipkoppe and also in the Bokkeveld and Hantam Mountains. **Flowers June to July.**

ANNELISE LE ROUX & NOEL VAN ROOYEN

Aloe komaggasensis
(formerly *Aloe striata* subsp. *komaggasensis*)
Komaggas aloe
ASPHODELACEAE   endemic

A succulent up to 70cm high. Usually stemless and lying on the ground, but very old plants sometimes form a short, creeping stem covered with persistent dried leaves. The dull grey-green, fleshy leaves are flat, broad and distinctly boat-shaped with faint longitudinal lines; the margin is spineless and yellow to orange. Tubular, yellow or orange flowers with a slight basal swelling are borne loosely together at the end of a branched stalk.

Found in stony soil on slopes in the Namakwaland Klipkoppe. **Flowers December to January.**

ANNELISE LE ROUX

Aloe krapohliana
Krapohl's aloe, Krapohl-se-aalwyn
ASPHODELACEAE

A succulent up to 20cm high. Usually with a single rosette up to 20cm in diameter. The fleshy, oblong leaves, up to 20cm long and 6cm wide at the base, lack spots or spines on the upper and lower surfaces and are grey-green with greyish-brown transverse bands; the margin has minute white teeth. Tubular, dull red flowers, about 3.5cm long and with greenish-yellow tips, are borne on stout, unbranched stalks.

Found on sandy flats and rocky slopes, often in quartz patches, throughout Namaqualand and the western part of Bushmanland.
Flowers June to August.

ANNELISE LE ROUX

Aloe melanacantha
goree, kleinbergaalwyn
ASPHODELACEAE

A succulent up to 30cm high, growing as single rosettes or more often in groups of up to 10 or more rosettes. The firm, fleshy, narrow, triangular, brownish-green leaves are about 20cm long and 4cm wide at the base, curving gracefully up- and inwards and giving the rosettes a neat, rounded appearance; the margin and midrib have black thorns up to 1cm long, the thorns sometimes relatively short with a whitish colour towards the base of the leaves, but distinctly black on the upper parts of the leaves. Tubular, bright red flowers that turn yellow with age are borne on a single erect stalk.

Found in rocky, mountainous areas in Namaqualand and also in southern Namibia.
Flowers May to June.

HEATHER BURGER

Aloe pearsonii
Pearson's aloe, Pearson-se-aalwyn

ASPHODELACEAE   endemic

A succulent shrub up to 1m high and 2m in diameter. The many, erect stems bear leaves for most of their length, branching occurring from the base of the plant. The almost triangular, fleshy leaves are dull bluish-green and often become red, especially in times of drought; they are arranged in neat vertical rows and curve backwards, giving a distinctive appearance. Red to orange-red or yellow flowers are borne sparsely at the tip of an erect stalk. Eaten by goats in times of severe drought.

Found on rocky mountain slopes in the Richtersveld.
Flowers December to January.

GONIALOE
There are three species of *Gonialoe* in southern Africa of which one occurs in Namaqualand.

Gonialoe variegata
variegated aloe, kanniedood, bontaalwyn

ASPHODELACEAE

A stemless succulent up to 30cm high. The fleshy, green or brownish, mottled leaves have a characteristic ridge or keel along the lower surface and lack spines or prickles; the margin has closely spaced, small teeth along the white, horny edge. Relatively large, hanging flowers varying from dull pink to red, rarely yellow, are borne on a branched stalk.

Found in loamy soil, occasionally in granitic soil, in Namaqualand and also in the Northern, Western and Eastern Cape, Free State and Namibia.
Flowers July to September.

Gasteria pillansii
ASPHODELACEAE

A stemless succulent up to 15cm high. The fleshy, dark, mottled, blunt, oblong leaves are arranged in 2 opposite rows and bend backwards; the surface is rough, with a white, toothed margin. Tubular, pink and pale green flowers, facing towards the same side of the stalk, hang down and are slightly swollen at the base; the stamens are slightly longer than the flower tube.

Found on south- and east-facing slopes in Namaqualand, on the Bokkeveld Escarpment and in the Tankwa Karoo, also in southern Namibia.
Flowers November to April.

PIETER VAN WYK

Haworthia arachnoidea
spinnekopbolletjie
ASPHODELACEAE   endemic

A perennial succulent up to 3.5cm high, although the plant is usually sunk into the ground. The uniform, plain green leaves are covered with many stiff, white hairs on the margin and keel, resembling a spider web. White flowers, tinged greyish-green, are borne on a slender stalk.

Found in the shelter of small bushes in rocky areas in Namaqualand.
Flowers November to March.

CARLO VAN TONDER

JEAN SMITH

CRINUM
There are 20 species of *Crinum* in southern Africa of which one occurs in Namaqualand.

Crinum variabile
river lily, rivierlelie
AMARYLLIDACEAE

A deciduous perennial herb up to 50cm high, with a bulb. The channelled, suberect to recurved leaves vary in length and width. Large, heavily scented, trumpet-shaped flowers are borne at the tip of an erect stalk; they are white to pale pink with deeper pink or green keels and become deeper pink with age; the white or pink stamens bend upwards.

Found along the edge of perennial rivers or in seasonal streambeds, in Namaqualand and also in the Bokkeveld and Roggeveld Mountains.
Flowers January to May.

ANNELISE LE ROUX

CROSSYNE
There are two species of *Crossyne* in southern Africa of which one occurs in Namaqualand.

Crossyne flava
parasol lily, sambreelblom
AMARYLLIDACEAE

A deciduous perennial herb up to 45cm high, with a deep-seated bulb. The 4–6 prostrate, broadly strap-shaped leaves, 5–10cm wide, have red, angular speckles on the undersurface; the margin is fringed with long, white, stiff hairs or straw-coloured bristles. Many small, asymmetrical, pale yellowish-green flowers about 1–1.5cm long are borne on long, rigid, widely spreading stalks in a round head 20–40cm across. The plant flowers before the leaves emerge.

Found in heavy loamy, granitic or sandstone-derived soil on flats or rocky slopes in Namaqualand and southwards to the Cederberg.
Flowers March to May.

STRUMARIA
There are 28 species of *Strumaria* in southern Africa
of which 10 occur in Namaqualand.

Strumaria truncata
bruidsklokkies

AMARYLLIDACEAE

A deciduous perennial herb up to 30cm
high, with a bulb. The 2–6 green, strap-
shaped leaves are usually twisted 1–3
times and arranged in a fan; they are
sheathed basally by an inflated, wine-red,
bladeless leaf. Funnel-shaped, scented,
white to pink, hanging flowers are borne
in clusters at the end of a long, often
bent stalk; the stamens are longer than
the petals. The plant flowers before the
leaves emerge.

Found in different types of soil, always
in fairly open sites, in Namaqualand and
also in the Bokkeveld Mountains.
Flowers April to June.

ANNELISE LE ROUX

HESSEA
There are 14 species of *Hessea* in southern Africa of
which eight occur in Namaqualand.

Hessea breviflora
matches, vuurhoutjies

AMARYLLIDACEAE

A deciduous perennial herb up to 20cm
high, with a bulb. The 2 emerging
leaves, enclosed by a reddened sheath,
are narrowly strap-shaped, hairless and
curved. Star- to widely funnel-shaped,
pink flowers are borne in small clusters at
the tips of erect stalks; the dark wine-red
anthers resemble matches, hence the
common name.

Found in loamy or sandy soil among
rocks in Namaqualand, extending
southwards to Hopefield in the
Western Cape.
Flowers April to May.

IVAN VAN NIEKERK

AMARYLLIDACEAE **107**

ANNELISE LE ROUX & ZELDA WAHL

There are about 20 species of *Brunsvigia* in southern Africa of which six occur in Namaqualand.

Brunsvigia bosmaniae
March lily, Maartblom, kandelaar
AMARYLLIDACEAE

A deciduous perennial herb up to 30cm high, with a bulb. The 6–8 broadly strap-shaped, leathery leaves, up to 12cm across, are pressed to the ground and have a rough upper surface; they appear in winter after the plant has flowered and set seed. Some 40–70 bright to pale pink flowers, up to 4.5cm long and with deeper pink veins, are borne in a round cluster at the end of a fleshy stalk; the outer stamens are about half the length of the inner ones.

Found on exposed flats in Namaqualand and the Bokkeveld Mountains, extending southwards into the Western Cape and northwards into Namibia.
Flowers March to April.

IVAN VAN NIEKERK & ANNELISE LE ROUX

Brunsvigia herrei
March lily, Maartblom, kandelaar
AMARYLLIDACEAE

A deciduous perennial herb up to 45cm high, with a bulb visible at the surface. The 6 or 7 suberect to spreading leaves are leathery and dull blue-green; the margin is red; the leaves appear in winter after the plant has flowered and set seed. Up to 40 delicate pink flowers with deeper pink veins are borne in a round cluster at the end of a fleshy stalk; the petals are up to 5.5cm long; the stamens are arranged in 2 whorls, the outer whorl half the length of the inner.

Found on rocky flats in mountainous areas of the Richtersveld and the Namakwaland Klipkoppe, extending into Namibia.
Flowers March to April.

108 AMARYLLIDACEAE

Brunsvigia namaquana
dwarf March lily, dwerg-Maartblom
AMARYLLIDACEAE

A deciduous perennial herb up to 15cm
high, with a bulb. The 2–4 broadly oblong
leaves lie flat on the ground; the upper
surface has straw-coloured bristles on
a thickened base; the leaves appear in
winter after the plant has flowered and
set seed. The 4–8 pink flowers are borne
in a round cluster at the end of a slightly
fleshy stalk; the petals bend upwards; the
stamens and style are longer than the
petals and usually bend downwards, with
the ends curving upwards.

   Found on sandy flats or rocky slopes
in the transition zone between winter-
and summer-rainfall areas in the eastern
Namakwaland Klipkoppe, extending
into Namibia.
Flowers November to March.

HEATHER BURGER

Brunsvigia orientalis
March lily, Maartblom
AMARYLLIDACEAE

A deciduous perennial herb up to 40cm
high, with a bulb. The 4–6 broadly strap-
shaped leaves lie flat on the ground; the
upper surface is velvety; the margin is
often fringed; the leaves appear in winter
after the plant has flowered and set seed.
The 20–40 red flowers, with unequal
petals that curve upwards, are borne in a
round cluster at the end of a fleshy stalk.

   Found on sandy flats, either in
riverbeds along the Coastal Plain
or on windblown dunes inland, in
Namaqualand, extending southwards
into the Western Cape.
Flowers February to March.

RIAAN DE VILLIERS

AMARYLLIDACEAE **109**

IVAN VAN NIEKERK

GETHYLLIS
There are about 30 species of *Gethyllis* in southern Africa of which 15 occur in Namaqualand.

Gethyllis britteniana
kukumakranka
AMARYLLIDACEAE

A deciduous perennial herb up to 20cm high, with a bulb. The erect, spiralled leaves are enclosed by a maroon and white mottled sheath at the base; the leaves appear in winter after the plant has flowered and set seed. The large, cup-shaped flowers are white to pink and have 30–60 stamens. Erect, cylindrical, fleshy, yellow fruits emerge from the ground after the flowers have died. The fruits are considered a delicacy and are also eaten by animals.

Found in sandy or rocky areas in Namaqualand, extending southwards to Clanwilliam and Darling.
Flowers October to March.

MARYNA KOHRS & PIETER VAN WYK

Gethyllis grandiflora
kukumakranka
AMARYLLIDACEAE   endemic

A deciduous perennial herb up to 30cm high, with a bulb. The 50 or more densely tufted, suberect, slightly twisted leaves are enclosed by a spotted basal sheath; the margin is occasionally sparsely hairy; the leaves appear in winter after the plant has flowered and set seed. The large white flowers are deeply cup-shaped and slightly fleshy; about 40 stamens are arranged in 6 clusters. Large, slightly fleshy, pale yellow to whitish fruits appear in summer after the flowers have died, but usually before the leaves emerge.

Found in sandy soil on rocky mountain slopes in the Richtersveld and the northern Namakwaland Klipkoppe.
Flowers October to December.

Gethyllis linearis
kukumakranka
AMARYLLIDACEAE

A deciduous perennial herb up to 10cm
high, with a bulb. The 5–10 leaves are
blue-green, narrowly strap-shaped,
spreading to suberect, tightly coiled,
mostly hairless and enclosed by a spotted
basal sheath; the leaves appear in winter
after the plant has flowered and set seed.
The trumpet-shaped flowers are white to
pale pink; there are 6 stamens; the style is
curved. Fleshy, yellowish fruits appear in
summer after the flowers have died, but
before the leaves emerge.
     Found on sandy flats in Namaqualand
and the Bokkeveld Mountains.
Flowers October to November.

ANNELISE LE ROUX

Gethyllis namaquensis
kukumakranka
AMARYLLIDACEAE

A deciduous perennial herb up to 30cm
high, with a bulb. The laxly spiralled leaves
are flat, densely tufted and enclosed by
a prominent, inflated, basal sheath; the
leaves appear in winter after the plant
has flowered and set seed. The large,
cup-shaped flowers are white or lilac; the
filaments of the 6 stamens are joined at
the base. Large, fleshy, pale yellow fruits
appear in summer after the flowers have
died, but usually before the leaves emerge.
     Found in coastal sand and on
mountain slopes in the Richtersveld,
extending into Namibia.
Flowers November to December.

MARYNA KOHRS

HAEMANTHUS
There are 22 species of *Haemanthus* in southern Africa of which nine occur in Namaqualand.

Haemanthus amarylloides subsp. *polyanthus*
April fool, Maartblom, skeerkwas
AMARYLLIDACEAE

A deciduous perennial herb up to 25cm high, with a bulb. The 2 or rarely 3 erect, soft-textured, hairless leaves are up to 30cm long and 2.5–5cm across; the margin, smooth or with a short membranous fringe, is sometimes thickened, occasionally outlined with yellowish-green or red; the leaves appear in winter after the plant has flowered and set seed. Many small, pale to dark pink flowers are borne in a cluster at the end of a stalk.

Found in sandy soils in the Namakwaland Klipkoppe and also in the Bokkeveld Mountains, Giftberg and Cederberg. **Flowers February to April.**

Haemanthus coccineus
April fool, Maartblom, skeerkwas
AMARYLLIDACEAE

A deciduous perennial herb up to 40cm high, with a bulb. The 2 or rarely 3 narrowly or broadly tongue-shaped leaves usually have reddish-brown or dark green bars on the underside; the margin is often rolled inwards; the leaves appear in winter after the plant has flowered and set seed. Many narrowly funnel-shaped, coral to scarlet flowers are borne in a cluster at the end of a spotted stalk; 6–9 overlapping, stiff, fleshy, scarlet bracts enclose the flower cluster.

Found in shaded kloofs and rock crevices and in flat, open areas in the shelter of shrubs, in Namaqualand, extending southwards to Port Elizabeth and northwards into Namibia. **Flowers February to April.**

Haemanthus crispus
April fool, skeerkwas
AMARYLLIDACEAE

A deciduous perennial herb up to 15cm high. The suberect to spreading, strap-shaped, channelled leaves are up to 2cm across; the undersurfaces are blotched or lined with dark green and reddish-brown; the margin is outlined with red and very wavy; the leaves appear in winter after the plant has flowered and set seed. The 6–25 tightly packed, narrowly funnel-shaped, scarlet or rarely pink flowers are borne on a red stalk; 4 or 5 blunt, pink to scarlet, waxy bracts enclose the flower cluster.

Found on rocky lower slopes in Namaqualand, extending southwards into the Olifants River Valley.
Flowers February to April.

ANNELISE LE ROUX

Haemanthus pubescens
April fool, skeerkwas
AMARYLLIDACEAE

A deciduous perennial herb up to 35cm high, with a bulb. The 2 strap-shaped to oblong leaves are bent backwards or lie flat on the ground; they are slightly hairy, occasionally smooth, and up to 4.5cm across; the margin is always softly fringed with hairs; the leaves appear in winter after the plant has flowered and set seed. The 4–60 narrowly funnel-shaped, scarlet or rarely pink flowers are tightly packed; 4 or 5 stiff, waxy, fleshy, red or rarely pink bracts enclose and overtop the flower cluster.

Found in loose, deep sand on the Coastal Plain and the Knersvlakte, extending southwards to Klawer and Graafwater and northwards into Namibia.
Flowers March to April.

ANNELISE LE ROUX

ASPARAGUS
There are about 85 species of *Asparagus* in southern Africa of which 17 occur in Namaqualand.

Asparagus asparagoides
ASPARAGACEAE

A deciduous, spineless, sprawling or scrambling shrub up to 1m high, with rhizomes. The many green, wiry branches are twisted and smooth or ridged. The shiny, green, pointed leaves are flat or folded and curved. The white flowers hang down, but the petals fold backwards; the erect stamens form a column. The round fruits are red when ripe.

Found in shady areas in the Namakwaland Klipkoppe and in other mountainous areas, extending southeastwards towards Port Elizabeth, from there to tropical Africa and to warm parts of Europe; it is also naturalised in Australia.
Flowers July to September.

Asparagus capensis
katdoring, wag-'n-bietjie
ASPARAGACEAE

A deciduous, erect to spreading, dense, spiny shrub up to 50cm high, with rhizomes. Branches have straight and curved spines in clusters of 3; the central spine is the longest. The small, grey-green leaves occur in clusters of 5 and are usually hairy. The white, stalkless flowers have spreading petals and stamens. Berries are about 4mm in diameter and turn red when mature.

Found on sandy flats or rocky slopes in Namaqualand and also in the Western Cape, Eastern Cape and Namibia.
Flowers April to August.

Asparagus juniperoides
cat's tail, katstert
ASPARAGACEAE

An evergreen, spineless shrub up to 50cm high, with rhizomes. This shrub has a shape similar to a cat's tail. The solitary, long, thin leaves overlap closely; the margin is fringed, mostly at the tips. The erect flowers are white; the lower petals are hairy.

Found on sandy flats or rocky slopes in Namaqualand, extending southwards to Clanwilliam and northwards into Namibia. **Flowers April to May.**

PIETER VAN WYK

Asparagus undulatus
ASPARAGACEAE

An evergreen, spineless, rigid shrub up to 40cm high, with rhizomes. The long, spreading, ridged branches bear regularly spaced leaves. The leathery, blue-green, sword-shaped, ribbed leaves are sharply pointed and fold inwards. The flowers hang down, but the petals fold backwards; the erect stamens form a column. The round fruits are red when ripe.

Found on sandy flats along the Coastal Plain and on the Knersvlakte, extending southwards to the Cape Peninsula and northwards into Namibia. **Flowers July to October.**

ANNELISE LE ROUX

ANNELISE LE ROUX

ERIOSPERMUM
There are about 100 species of *Eriospermum* in
southern Africa of which 40 occur in Namaqualand.

Eriospermum arachnoideum
RUSCACEAE   endemic

A deciduous perennial herb up to 5cm
high, with a tuber. The single, heart-
shaped leaf lies flat on the ground and is
covered with cobweb-like hairs borne on
raised pustules; the margin is transparent;
the leaves are produced in winter and die
off before the flowers are produced in
summer when they are seldom noticed.
Up to 7 white flowers are borne on an
erect, red-streaked stalk up to 10cm long;
the narrowly elliptic outer petals spread
outwards and the spoon-shaped inner
petals stand erect and have a pinched tip;
the anthers are pale mauve.
  Found in crevices of dolomite outcrops
on the Knersvlakte.
Flowers February to March.

NICK HELME

Eriospermum calcareum
RUSCACEAE   endemic

A deciduous perennial herb up to 10cm
high, with a tuber. The single, slightly
leathery leaf is heart-shaped and hairless,
or sparsely hairy near the edges; the
brown leaf margin is thickened and
minutely scalloped; the leaves are
produced in winter and die off before the
flowers are produced in summer when
they are seldom noticed. Up to 14 white
flowers are borne on an erect stalk up
to 10cm long; the narrowly elliptic outer
petals spread outwards and the spoon-
shaped inner petals stand erect.
  Found on dolomite outcrops,
occasionally in red sand deposits,
on the Knersvlakte.
Flowers February to March.

Eriospermum descendens
RUSCACEAE   endemic

A deciduous perennial herb up to 15cm
high, with a tuber. Plants are often found
in clumps. The single, slightly fleshy leaf
lies flat on the ground and is heart-shaped,
with basal lobes that occasionally overlap;
the margin is red; the leaves are produced
in winter and die off before the flowers are
produced in summer when they are seldom
noticed. Up to 30 star-shaped, cream-
coloured flowers are borne on an erect
stalk up to 10cm long; the white filaments
are broad and flat with relatively small,
yellow anthers; the stamens resemble an
inner whorl of petals.

　　Found in sandy or loamy soil in the
Namakwaland Klipkoppe and on the
Knersvlakte.
Flowers February to April.

Eriospermum paradoxum
haasklossie
RUSCACEAE

A deciduous perennial herb up to 10cm
high, with a tuber. The single small, hairy,
green leaf is tinged purple; an erect,
white, woolly, finely dissected appendage
emerges from the centre of the leaf; the
leaves are produced in winter and die
off before the flowers are produced in
summer when they are seldom noticed. Up
to 40 star-shaped, sweetly scented, white
flowers are borne on an erect stalk up to
9cm long.

　　Found in sandy or loamy soil in
Namaqualand, extending southeastwards
to Grahamstown.
Flowers April to May.

RUSCACEAE **117**

ANNELISE LE ROUX

CHLOROPHYTUM
There are about 40 species of *Chlorophytum* in southern Africa of which seven occur in Namaqualand.

Chlorophytum crassinerve

AGAVACEAE   endemic

A deciduous perennial herb up to 40cm high, with rhizomes. The 4 or 5 sword-shaped, leathery basal leaves are prominently and closely ribbed, up to 14cm long and 2cm across and grow in a rosette; the thickened margin is red. White flowers, about 3cm in diameter and with red keels, are borne on an unbranched, spotted stalk.

Found in stony soil in the Namakwaland Klipkoppe.
Flowers August to October.

ZELDA WAHL

Chlorophytum undulatum

AGAVACEAE

A deciduous perennial herb up to 50cm high, with rhizomes. The 3–9 linear to sword-shaped leaves are 5–20cm long and 2–10mm across and usually grow in a rosette; the margin is straight or crisped and minutely fringed. White flowers about 1.5–3cm in diameter are closely spaced on an unbranched stalk.

Found in stony areas in Namaqualand, the Bokkeveld and Roggeveld Mountains and also in the Western Cape.
Flowers July to October.

Chlorophytum viscosum
AGAVACEAE

A deciduous perennial herb up to 60cm high, with rhizomes. The entire plant is covered with glands borne on minute stalks. The ribbed, leathery, linear to sword-shaped leaves are up to 30cm long and usually grow in a rosette. The white flowers have spreading petals; the filaments are rough with minute hairs.

Found on sandy flats or rocky slopes in Namaqualand, extending southwards to Piketberg and northwards into Namibia. **Flowers June to October.**

PIETER VAN WYK

ALBUCA
There are about 110 species of *Albuca* in southern Africa of which 39 occur in Namaqualand.

Albuca canadensis
(formerly *A. maxima*)
slymstok, kamiemie
HYACINTHACEAE

A deciduous perennial herb up to 1m high, with a bulb. The 5 or 6 bluish-green, slightly fleshy leaves, up to 60cm long and 5cm across at the base, are sword-shaped and tapered towards the tip. White and green flowers up to 2.5cm long are borne loosely on an unbranched stalk; the outer stamens are sterile.

Found on rocky slopes, often on roadsides, in Namaqualand, extending into the Western Cape. **Flowers August to October.**

ZELDA WAHL

ANNELISE LE ROUX

Albuca ciliaris
HYACINTHACEAE

A deciduous perennial herb up to 20cm high, with a bulb. The 5–20 bluish-green leaves are narrowly oblong and often longitudinally twisted; the margin is hairy. Dull greenish, hanging flowers up to 2.5cm long are borne loosely on an unbranched stalk; the outer stamens are sterile.

Found on rocky flats and slopes in Namaqualand, extending southwards to Clanwilliam.
Flowers August to October.

ZELDA WAHL

Albuca consanguinea
(formerly *Ornithogalum polyphyllum*)
HYACINTHACEAE

A deciduous perennial herb up to 60cm high, with a bulb. The erect, basal leaves are channelled, linear to almost cylindrical and up to 25cm long and 8mm across, the tips drying out and becoming loosely coiled with age. Numerous fragrant, white or yellow flowers, up to 3cm in diameter and with green stripes, are borne loosely on a stalk.

Found on rocky slopes in the Nama-kwaland Klipkoppe, on the Knersvlakte and the Hantam Mountains, extending southwards to Tulbagh.
Flowers August to September.

Albuca hallii
HYACINTHACEAE

A deciduous perennial herb up to 20cm high, with a bulb. The 3–6 slightly fleshy leaves are covered with glands and shaped like a corkscrew towards the tip; the outer surface is rounded and the inner surface flat. The hanging yellow flowers with broad, green midribs are up to 1.5cm long.

Found on rocky slopes in Namaqualand, the Hantam Mountains and the Little Karoo, also in Namibia. **Flowers March to May.**

PIETER VAN WYK

Albuca leucantha
HYACINTHACEAE

A deciduous perennial herb up to 50cm high, with a bulb. The basal leaves are linear and up to 30cm long and 1.5cm across; the margin is rolled inwards. The hanging white flowers have broad, green midribs; the outer stamens are sterile.

Found in rocky areas in Namaqualand, extending into Namibia. **Flowers August to October.**

ANNELISE LE ROUX

Albuca longipes
HYACINTHACEAE

A deciduous perennial herb up to 50cm high, with a bulb. The 1–6 linear, channelled leaves are often dry at flowering. White flowers, up to 2cm long and with broad, green midribs, are borne on long, erect to suberect stalks of equal length.

Found in sandy soil on stony flats or slopes in Namaqualand, extending into the Western Cape.
Flowers September to November.

PIETER VAN WYK

Albuca secunda
(formerly *Ornithogalum secundum*)
HYACINTHACEAE

A deciduous perennial herb up to 35cm high, with a bulb. The strap-shaped leaves, usually arranged in a rosette lying flat on the ground, are withered or absent at flowering; the margin is minutely fringed and translucent. Many yellow flowers with broad green midribs are borne on an erect stalk; the oblong petals are spreading.

Found in sandy soil in Namaqualand, extending into the Western Cape.
Flowers September to October.

ANNELISE LE ROUX & ZELDA WAHL

Albuca spiralis

HYACINTHACEAE

A deciduous perennial herb up to 40cm high, with a bulb. The 5–10 basal leaves, up to 30cm long and 4mm across, are channelled, often spirally twisted or coiled, and are covered with glandular hairs. The 2–4 pale yellow-green, hanging flowers, up to 1.8cm long and with green stripes, are sparsely arranged on an erect stalk; all stamens are fertile.

Found on sandy, stony slopes in the Namakwaland Klipkoppe and on the Knersvlakte, extending southwards to the Cape Peninsula.

Flowers August to October.

ZELDA WAHL

Albuca suaveolens

(formerly *Ornithogalum suaveolens*)

HYACINTHACEAE

A deciduous perennial herb up to 60cm high, with a bulb. The 2–4 bluish-green, basal leaves, up to 25cm long and 2cm across, are linear and slightly channelled above. The few to many sweetly scented flowers are yellow with green median stripes and up to 3cm in diameter. The flowers open only in the morning.

Found on sandy flats or rocky slopes in Namaqualand, extending southwards to the Cape Peninsula.

Flowers August to October.

ZELDA WAHL

HYACINTHACEAE **123**

ORNITHOGALUM
There are about 100 species of *Ornithogalum* in southern Africa of which 22 occur in Namaqualand.

Ornithogalum falcatum
HYACINTHACEAE

A deciduous perennial herb up to 8cm high, with a bulb. The single, slightly fleshy, blue-green leaf is sickle-shaped and up to 6mm across; the margin is thickened and reddish. The bracts, located at the base of the flowers, are membranous and egg-shaped at the base, with lacerated margins; the distal half is sharply pointed. The flowers are yellow, tinged purple; the filaments are joined at the base and about 1.5mm long; the style is 1.5mm long and 3-branched.

Found in sandy soil on the Coastal Plain of Namaqualand, near Alexander Bay. **Flowers May to July.**

Ornithogalum filicaule
HYACINTHACEAE   endemic

A deciduous perennial herb up to 10cm high, with a bulb. The 2 oblong leaves lie flat on the ground. The green, slightly fleshy bracts are at the base of the flowers. Pale green, slightly fleshy flowers point upwards and are borne on long stalks; the filaments are joined at the base and are about 1.5mm long; the style is up to 1.5mm long and 3-branched.

Found on stony slopes in the Namaqualand Klipkoppe. **Flowers May to July.**

Ornithogalum hispidum
HYACINTHACEAE

A deciduous perennial herb up to 40cm high, with a bulb. The 3–6 leaves clasp the stem and one another and are sparsely to densely covered with hairs. The base of the long flowering stalk, sheathed by the leaves, is straight or slightly curved. The white or cream-coloured, star-shaped flowers have inner filaments that are broadened and flattened at the base.

Found in loamy soils, sometimes in white quartz fields, in Namaqualand, extending southwards to the Cape Peninsula and into the Little Karoo and northwards into Namibia.
Flowers August to October.

ANNELISE LE ROUX

Ornithogalum maculatum
chinkerinchee, rooi-dirkie, nagslangetjie
HYACINTHACEAE

A deciduous perennial herb up to 50cm high, with a bulb. The 2–5 basal leaves are blue-green, linear to elliptic, erect or suberect and up 15cm long. Bracts on the flowering stalk are papery. The 1–8 flowers are yellow, orange or orange-red; the 3 outer petals are up to 3cm long and usually have a black or yellow marking at the tip.

Found in shallow rock pockets filled with sandy soil, in the Namakwaland Klipkoppe, on the Knersvlakte and also in the Bokkeveld Mountains, the Cederberg, the Tankwa Karoo and the Little Karoo.
Flowers September to October.

ZELDA WAHL

Ornithogalum pruinosum
chinkerinchee, wit-dirkie
HYACINTHACEAE

A deciduous perennial herb up to 50cm high, with a bulb. The 4–6 basal leaves are pale bluish-green, oblong to sword-shaped and up to 1.7cm long and 3cm across. Numerous white flowers, about 3cm in diameter and with faint green stripes, are densely arranged on a stalk.

Found on rocky slopes in Namaqualand and in the Hantam Mountains. **Flowers July to October.**

ANNELISE LE ROUX

Ornithogalum rupestre
(formerly *O. multifolium*)
rock chinkerinchee, klip-dirkie
HYACINTHACEAE

A deciduous perennial herb up to 10cm high, with a bulb. The greyish-green leaves are slightly fleshy, cylindrical and often slightly twisted. The scented cup- to star-shaped flowers are yellow or orange; the stamens are half the length of the petals; the filaments are broader towards the base.

Found in shallow pockets of sandy soil on rocky outcrops in the Richtersveld, the Namakwaland Klipkoppe and on the Knersvlakte, extending southwards to Mamre in the Western Cape. **Flowers August to October.**

ANNELISE LE ROUX

Ornithogalum thyrsoides
chinkerinchee
HYACINTHACEAE

A deciduous perennial herb up to 50cm high, with a bulb. The leaves are suberect, sword-shaped and sometimes dry at flowering; the margin is hairy. The broad, white, shiny bracts are located at the base of the flowers. The white, star- or bowl-shaped flowers usually have a dark brown or green centre; the 3 inner stamens have a broadened and flattened lower half and curve over the ovary. Extensively cultivated for the cut-flower trade. Poisonous to horses and cattle.

Found in seasonally damp areas, on disturbed land, sandy flats or lower mountain slopes, in the Namakwaland Klipkoppe and on the Knersvlakte, extending southwards to Bredasdorp. **Flowers September to November.**

ANNELISE LE ROUX

Ornithogalum xanthochlorum
slangkop
HYACINTHACEAE

A deciduous perennial herb up to 60cm high, with a bulb. The 10–12 basal leaves are erect to spreading, strap-shaped and up to 30cm long and 3.5cm across; the margin is semitransparent. The bracts are longer than the flowers. Greenish flowers up to 2.5cm in diameter are borne close together on a stalk.

Found on sandy flats in Namaqualand, the Hantam Mountains and the Tankwa Karoo. **Flowers August to September.**

ZELDA WAHL

DAVE GWYNNE-EVANS

DIPCADI
There are about 15 species of *Dipcadi* in southern Africa of which two occur in Namaqualand.

Dipcadi brevifolium
HYACINTHACEAE

A deciduous perennial herb up to 40cm high, with a bulb. The 2–4 leaves are cylindrical or channelled, straight or recoiled and often form loose spirals when young; the leaves can be present or absent at flowering time. The flowers vary from brown to green, greenish-yellow to cream-coloured and hang down from the stalk; the petals are fused in the basal half; the outer petals are narrower and slightly longer than the inner ones and bend backwards from halfway up, while the inner petals form a tube with only the tips spreading.

Found on rocky flats or slopes in Namaqualand, extending through western and central South Africa and Namibia. **Flowers August to April.**

PIETER VAN WYK

Dipcadi crispum
krul-ui
HYACINTHACEAE

A deciduous perennial herb up to 30cm high, with a bulb. The grey-green leaves are narrowly sword-shaped, straight or spirally twisted and up to 1.6cm across; the margin is straight or crisped and sparsely or densely fringed with long white hairs. The sweetly scented flowers are brown to orange, up to 2cm long and hang down from the stalk; the petals are fused at the base into a bulbous tube for about one third of their length; the outer petals are slightly longer and narrower than the inner ones.

Found in rocky areas in Namaqualand, the Western Cape and Namibia. **Flowers April to December.**

EUCOMIS
There are 11 species of *Eucomis* in southern Africa of which one occurs in Namaqualand.

Eucomis regia
pineapple lily, pynappellelie
HYACINTHACEAE

A deciduous perennial herb up to 25cm high, with a bulb. The basal leaves are broadly spoon- to egg-shaped, lie flat on the ground and are sometimes speckled with purple towards the base of the undersurface. Scented green flowers, tightly clustered on a thick stalk, are overtopped by a large conspicuous tuft of green bracts.

Found in shade under bushes, along seasonal streams and on south-facing rocky slopes, in the Namakwaland Klipkoppe and the Bokkeveld and Roggeveld Mountains, extending southwards to the Little Karoo. **Flowers July to September.**

ANNELISE LE ROUX

VELTHEIMIA
There are two species of *Veltheimia* in southern Africa of which one occurs in Namaqualand.

Veltheimia capensis
sandlelie, berglelie
HYACINTHACEAE

A deciduous perennial herb up to 40cm high, with a bulb. The 4–9 basal leaves are bluish-green, sword-shaped to oblong and up to 30cm long and 9cm across; the margin is often crisped or wavy. Many purplish-pink flowers, about 3cm long and with green tips, are borne close together on the upper third of the stalk. Fruits are inflated capsules.

Found on sandy flats and rocky slopes in Namaqualand and in dry areas of the Western Cape. **Flowers April to August.**

VERONICA ESTERHUIZEN & ZELDA WAHL

HYACINTHACEAE **129**

Massonia bifolia
(formerly *Whiteheadia bifolia*)
elephant's ear, olifantsoor
HYACINTHACEAE

A deciduous perennial herb up to 20cm high, with a bulb. The 2 opposite leaves are egg-shaped, up to 30cm long and 20cm across, and lie flat on the ground. Cup-shaped greenish flowers about 1cm in diameter are closely packed on a stalk between bracts longer than the flowers.

Found in the shade of rocks on hills in Namaqualand, the Bokkeveld and Pakhuis Mountains and in Namibia.
Flowers July to September.

VERONICA ESTERHUIZEN

Massonia depressa
suikerkannetjie, botterkannetjie, ramblom, karring
HYACINTHACEAE

A deciduous perennial herb growing flat on the ground from a bulb. The 2 opposite leaves are egg-shaped, up to 17cm long and 15cm across. White flowers with reddish stamens are borne in a dense cluster with numerous reddish-margined, green bracts in the centre of the plant. The bulb of this plant can be eaten and the nectar of the flowers is very sweet.

Found on sandy flats in Namaqualand and elsewhere in southwestern South Africa.
Flowers June to August.

LACHENALIA
There are about 135 species of *Lachenalia* in southern
Africa of which 40 occur in Namaqualand.

Lachenalia anguinea
viooltjie
HYACINTHACEAE

A deciduous perennial herb up to 30cm
high, with a bulb. The single leaf, up to
25cm long, about 2cm across, is sword-
shaped and banded with cross-bars of
darker green on the back. The numerous
bell-shaped flowers, about 1cm long, are
white tipped with green; the inner petals
are scarcely longer than the outer ones;
the stamens are longer than the petals.

Found in sandy areas in Namaqua-
land, extending southwards into the
Western Cape.
Flowers July to September.

MARINDA KOEKEMOER

Lachenalia carnosa
viooltjie
HYACINTHACEAE   endemic

A deciduous perennial herb up to 20cm
high, with a bulb. The 2 egg- to sword-
shaped leaves, up to 8cm long, lie on the
ground and often have a few to many
purplish markings on the upper side;
the margin is brownish-maroon. The
numerous urn-shaped white flowers,
about 7mm long and with maroon tips,
are densely arranged on a swollen stalk;
the stamens are shorter than the petals.

Found in sandy areas in the Namakwaland Klipkoppe.
Flowers August to September.

ANNELISE LE ROUX

HYACINTHACEAE **131**

VERONICA ESTERHUIZEN

Lachenalia framesii
viooltjie
HYACINTHACEAE   endemic

A deciduous perennial herb up to 15cm
high, with a bulb. The 1 or 2 leaves clasp
the base of the plant; the margin is usually
undulate. Urn-shaped to oblong yellow
flowers, with brown or green swellings on
the outer petals, are borne sparsely on the
stalk; the inner petals have distinct purple-
mauve tips, are bent backwards and are
slightly longer than the outer ones; the
stamens are shorter than the petals.

Found in red, loamy soil on flats in
the Namakwaland Klipkoppe and on
the Knersvlakte.
Flowers July to August.

ANNELISE LE ROUX

Lachenalia glauca
viooltjie
HYACINTHACEAE

A deciduous perennial herb up to 25cm
high, with a bulb. The 1 or 2 leaves are
narrowly or broadly sword-shaped, blue-
green to pale green and clasp the base of
the plant; the margin is green or maroon,
flat, or slightly to moderately undulate.
The numerous hanging, bell-shaped
flowers are coconut-scented and up to
1cm long; the outer petals are grey-blue
at the base, shading to pale magenta
with darker keels and tips, while the
slightly protruding inner petals are also
pale magenta; the stamens, with pale
magenta or purple filaments, are longer
than the petals.

Found on damp, sandy flats or on flat
granite outcrops in the Namakwaland
Klipkoppe and also near Clanwilliam.
Flowers July to September.

Lachenalia hirta
viooltjie
HYACINTHACEAE

A deciduous perennial herb up to 25cm high, with a bulb. The single, linear leaf has long, stiff hairs and conspicuous bands of maroon on the lower surface; the margin is wavy or straight, with long, stiff hairs. Tubular flowers about 1cm long are borne on long stalks; the outer petals are pale blue to blue-grey at the base, with brown pouch-like swellings at the tips; the inner petals are longer and pale yellow with small brown marks.

Found in flat, sandy areas in the Namakwaland Klipkoppe, the Bokkeveld Mountains and the Cederberg, extending southwards to Cape Town.
Flowers August to September.

TESSA OLIVER

Lachenalia inconspicua
viooltjie
HYACINTHACEAE

A deciduous perennial herb up to 15cm high, with a bulb. The usually single blue-green leaf is leathery, sword- to egg-shaped, widely spreading and deeply channelled; it has irregularly scattered purplish-brown spots on the upper surface. The few to many flowers are oblong to bell-shaped and pale bluish- or greenish-white fading to dull brownish-purple; the outer petals, bent slightly backwards, are pale greenish-blue with darker blue markings at the base and dull purplish-brown or brownish-green swellings at the tip; the inner petals, protruding well beyond the outer ones, bend slightly backwards and are translucent white, with brownish-green keels.

Found in deep sand in the Namakwaland Klipkoppe and in Bushmanland.
Flowers July to August.

ANNELISE LE ROUX

Lachenalia minima
viooltjie
HYACINTHACEAE   endemic

A deciduous perennial herb up to 15cm high, with a bulb. The 2 sword- to egg-shaped, spreading leaves have depressed veins, with or without pustules, on the upper surface. The oblong bell-shaped flowers are pale yellow, changing to dull red as they mature; the outer petals have green swellings while the inner petal segments protrude slightly; the stamens are shorter than the petals.

Found in loamy soil on flats on the Knersvlakte.
Flowers June.

Lachenalia mutabilis
viooltjie
HYACINTHACEAE

A deciduous perennial herb up to 30cm high, with a bulb. The 1 or occasionally 2 leaves are sword-shaped, with or without maroon bands at the base of the outer surface; the margin is wavy or crisped. The tubular flowers are about 1.5cm long; the outer petals are blue to mauve and the inner ones dark yellow; the stamens are normally shorter than the petals; the aborted buds at the tip are magenta to blue.

Found in sandy or stony areas, often in clumps, in Namaqualand and the Bokkeveld Mountains, extending southwards into the Western Cape.
Flowers July to September.

Lachenalia namaquensis
viooltjie

HYACINTHACEAE   endemic

A deciduous perennial herb up to 20cm high, with a bulb. The 1 or 2 long, suberect or arched leaves are folded lengthwise. The narrow, urn-shaped flowers are about 1cm long; the outer petals are purplish to magenta with maroon tips, the upper inner petals are whitish and the lower inner petal is longer and deep magenta; the stamens are shorter than the petals.

Found in loamy soil on flats and rocky slopes in the Richtersveld, Namakwaland Klipkoppe and on the Coastal Plain. **Flowers August to October.**

ZELDA WAHL

Lachenalia obscura
viooltjie

HYACINTHACEAE

A deciduous perennial herb up to 30cm high, with a bulb. The 1 or 2, yellowish-green to dark green leaves have faint, depressed longitudinal veins on the upper surface; the lower surface is usually heavily banded with bright green merging into dull brownish-purple and magenta. Bell-shaped flowers are borne on a stalk blotched with pale to dark purple; the outer petals are oblong, pale yellowish-green to brownish-blue, or cream with pale blue speckles or solid blue at the base, with dull brown, purplish or green swellings and slightly recurved tips; the inner petals protrude well beyond the outer ones; stamens slightly longer than the petals.

Found in loamy soil on slopes in the Namakwaland Klipkoppe, the Hantam and Roggeveld Mountains and Little Karoo. **Flowers June to October.**

HYACINTHACEAE **135**

VERONICA ESTERHUIZEN

NICOLA BERGH

Lachenalia patula
viooltjie
HYACINTHACEAE   endemic

A deciduous perennial herb up to 10cm high, with a bulb. The 2 erect, very fleshy leaves are green, ageing to a dull or bright deep maroon. A few large, widely bell-shaped, white flowers, with maroon marks at the tips and keels of the petals, are borne at the ends of a maroon stalk; the stamens are shorter than the petals.

Found in shallow, loamy soil among white quartz pebbles on the Knersvlakte. **Flowers September to October.**

VERONICA ESTERHUIZEN

Lachenalia polypodantha
viooltjie
HYACINTHACEAE

A deciduous perennial herb up to 10cm high, with a bulb. The single, fleshy leaf, clasping the base of the plant, is sword-shaped to heart-shaped with shallow, depressed longitudinal grooves and can be unmarked or flushed with maroon speckles; the upper leaf surface is covered with star-shaped hairs. Bell-shaped flowers, with white petals with green or brown keels and tips, are borne on a maroon-purple to pale green or ivory-white stalk with or without minute maroon blotches; the stamens, with white or violet filaments, are much longer than the petals.

Found in deep sand in the Richtersveld, the Namakwaland Klipkoppe and also in Bushmanland. **Flowers August to September.**

Lachenalia punctata
sandkalossie, rooiviooltjie

HYACINTHACEAE

A deciduous perennial herb up to 20cm high, with a bulb. The erect, suberect or spreading leaves are plain to heavily marked with small to large purple blotches on the upper surface. Long, cylindrical, hanging flowers vary in colour from finely mottled bright pink to finely mottled ruby-red; the outer petals have yellowish-green or pinkish-red marks; the inner segments protrude conspicuously.

Found on sandy flats on the Coastal Plain and in the Namakwaland Klipkoppe, extending along the coast to Mossel Bay in the Western Cape.
Flowers March to July.

ANNELISE LE ROUX

Lachenalia pygmaea
viooltjie

HYACINTHACEAE

A deciduous perennial herb up to 5cm high, with a bulb. The 2 egg-shaped leaves lie flat on the ground and have distinctly depressed longitudinal grooves. Small, strongly almond-scented, erect, tubular flowers are borne in a congested cluster at ground level between the 2 leaves; the flower tubes are white and the pale to bright mauve petals are rolled back; the stamens are longer than the rolled back petals.

Found in large colonies on seasonally damp flats or in sandy pockets on flat, rocky outcrops, in the Namakwaland Klipkoppe and also near Bredasdorp in the Western Cape.
Flowers April to June.

ANNELISE LE ROUX

VERONICA ESTERHUIZEN

Lachenalia schlechteri
viooltjie
HYACINTHACEAE

A deciduous perennial herb up to 30cm high, with a bulb. The single linear leaf widens conspicuously in the lower portion, forming a loosely clasping base, usually distinctly banded with maroon and magenta. Long, cylindrical flowers are borne on a stalk that can be plain or lightly to heavily mottled with dark brownish-maroon; the flower tube is white with pale to deep blue or maroon blotches or bands; the lower inner petals are longer than the outer ones; the stamens are shorter than the petals.

Found in flat areas in the Namakwa-land Klipkoppe and also in the Olifants River Valley.
Flowers August to September.

CARLO VAN TONDER

Lachenalia trichophylla
viooltjie
HYACINTHACEAE

A deciduous perennial herb up to 20cm high, with a bulb. The single egg-shaped leaf is covered with erect star-shaped hairs of varying length on the upper surface; the undersurface is maroon. The tubular flowers vary in shades of pale yellow; the outer petals have green blotches and may be flushed pink, while the inner ones are longer and protrude; the stamens are shorter than the petals.

Found in sandy, stony soil in the Namakwaland Klipkoppe and also in the Bokkeveld, Cederberg and Hantam Mountains.
Flowers July to September.

Lachenalia undulata
viooltjie
HYACINTHACEAE

A deciduous perennial herb up to 20cm high, with a bulb. The 2 suberect to spreading, egg- to sword-shaped leaves can be plain or have irregular dark brown spots; the margin is slightly to distinctly thickened and usually undulate. Stalkless, oblong to bell-shaped, whitish flowers that turn red-brown as they mature, are borne horizontally on a stout stalk; the outer petals are dull white, tinged green and pale blue, with dark brown swellings, while the inner ones are white, protruding from the shorter outer petals.

Found on rocky flats or slopes in Namaqualand, the Bokkeveld Mountains and in Bushmanland.
Flowers May to June.

ANNELISE LE ROUX

Lachenalia valeriae
viooltjie
HYACINTHACEAE   endemic

A deciduous perennial herb up to 25cm high, with a bulb. The 2 suberect to spreading leaves are up to 3cm across and the upper surface is densely covered with small raised spots. Urn-shaped, stalkless flowers are borne on a sturdy, speckled, brownish-purple stalk; they have greenish-yellow to white outer petals with green swellings on the keels and at the tips; the inner petals are flushed bright magenta at the tips; the lower inner petal is longer than the others; the stamens are shorter than the petals.

Found singly or in large colonies, on deep, sandy flats or in pockets on granite or shale outcrops, in the Namakwaland Klipkoppe and on the Coastal Plain.
Flowers July to August.

PIETER VAN WYK

PIETER VAN WYK

Lachenalia violacea
viooltjie
HYACINTHACEAE

A deciduous perennial herb up to 25cm high, with a bulb. The 1 or 2 blue-green leaves, up to 10cm long, are sword-shaped and channelled above; both surfaces plain or heavily spotted with brownish-purple; margin thickened, reddish-brown and wavy or crisped. Hanging, bell-shaped, bluish-mauve flowers are borne on a sturdy, often inflated stalk that can be plain or blotched with purple; the outer petals are pale blue-green to white at the base, shading to pale magenta or purple with brown swellings; the inner petals, varying in shades of purple or violet, protrude slightly from the outer ones; stamens much longer than the petals.

Found on rocky slopes or in deep sand near seasonal streams in Namaqualand, extending southeastwards to Fraserburg. **Flowers July to September.**

JEAN SMITH

Lachenalia xerophila
viooltjie
HYACINTHACEAE

A deciduous perennial herb up to 25cm high, with a bulb. The 1 or occasionally 2 suberect leaves are blue-green to bright green and channelled; the margin is thickened, brownish, entire to markedly undulate or crisped. Numerous longish, bell-shaped flowers are borne on an erect to suberect, sturdy stalk that is conspicuously inflated at the base, gradually becoming less inflated towards the tip; the outer petals are white with brown keels and swellings at the tips; the inner petals are slightly longer than the outer ones; the stamens are longer than the petals.

Found singly, or in colonies, in deep, red sand on the Coastal Plain, extending through the Namakwaland Klipkoppe into Bushmanland.
Flowers July to September.

140   HYACINTHACEAE

WACHENDORFIA
There are four species of Wachendorfia in southern
Africa of which two occur in Namaqualand.

Wachendorfia multiflora
kleinrooikanol
HAEMODORACEAE

A deciduous perennial herb up to 30cm
high, with rhizomes. The softly hairy,
lance-shaped leaves, up to 36cm long and
2.5cm across, are usually longer than the
flowering stalk. Bracts at the base of the
flowers are herbaceous, green and erect.
The dull yellow flowers fade to brownish-
purple; the petals are narrow, 3–6mm
across; the stamens are two-thirds the
length of the petals.

Found in dry, sandy hollows on the
coast and on damp, rocky ledges in the
mountains of Namaqualand, extending
southwards to the Robertson area.
Flowers August to September.

HERGEN ERHARDT

Wachendorfia paniculata
rooikanol
HAEMODORACEAE

A deciduous perennial herb up to 60cm
high, with rhizomes. The narrowly lance-
shaped, erect or spreading leaves, up
to 3.5cm across, are usually hairy and
shorter than the flowering stalk. Bracts
at the base of the flowers are thin, dry
and membranous, not green, and often
recurved. The apricot, yellow or orange
flowers are slightly scented, 4–16mm
across; the petals are broad, 5–18mm
across; the stamens are from two-thirds to
three-quarters the length of the petals.

Found in loose, sandy soil on the
Knersvlakte and also in the Western and
Eastern Cape.
Flowers August to November.

RIAAN DE VILLIERS

FICINIA
There are about 70 species of *Ficinia* in southern Africa of which 11 occur in Namaqualand.

Ficinia indica
knoppiesbiesie
CYPERACEAE

A compact or loosely tufted perennial sedge up to 45cm high, with runners radiating from the base. The leaf sheaths are reddish; the margin is rough. The broadly ovoid flowerheads are terminal, compact and chestnut-brown.

Found in damp, sandy areas on the Coastal Plain and Knersvlakte, extending into the Western and Eastern Cape. **Flowers July to November.**

Ficinia nigrescens
swartkopbiesie
CYPERACEAE

A tufted, erect perennial sedge up to 40cm high. The leaf sheaths are reddish; the leaves are flattened or infolded and slightly curved. Light to dark glossy brown flowers are clustered together at the top of a stalk.

Found in dry, rocky areas in mountainous terrain in Namaqualand, extending into the Western and Eastern Cape. **Flowers May to October.**

There are about 40 species of *Isolepis* in southern Africa of which 14 occur in Namaqualand.

Isolepis incomtula
CYPERACEAE

A tufted annual herb up to 10cm high. The linear leaves are 2–8cm long and erect. The flowers are clustered together at the top of a thin, erect stalk.

Found in damp areas, usually on streambanks, in Namaqualand, extending into the Western and Eastern Cape. **Flowers July to October.**

ZELDA WAHL

WILLDENOWIA
There are 12 species of *Willdenowia* in southern Africa of which two occur in Namaqualand.

Willdenowia incurvata
sonkwasriet, besemriet

RESTIONACEAE

A tufted perennial reed with long, rigid, branched stems up to 1m high. The many male flowers are loosely grouped, while the female flowers (on separate plants) are borne singly on long stalks.

Found in dense to sparse colonies in red sand along the Coastal Plain of Namaqualand and into the Western Cape. **Flowers April to June.**

RIAAN DE VILLIERS

ZELDA WAHL

EHRHARTA
There are about 25 species of *Ehrharta* in southern
Africa of which nine occur in Namaqualand.

Ehrharta calycina
rooisaadgras, rooipootjie
POACEAE

A robust or weakly tufted perennial
grass up to 60cm high. The leaf blades
are flat or rolled; the margin is often
undulate. Flowers are borne sparsely
on a thin stalk. A very palatable grass
utilised by cattle more than by sheep. It
is mainly the new growth of the leaves
and stalks that is eaten, but at the end
of the growth period only the leaves,
not the stalks, are utilised.

Found in Namaqualand, extending
from Namibia to KwaZulu-Natal.
**Flowers July to December, but usually
in spring.**

ZELDA WAHL

Ehrharta longiflora
pêpê-gras
POACEAE

An annual grass with stalks up to 35cm
high. The flat leaves are usually about
5mm wide; the margin is undulate. Long
flowers up to 2.5cm long are borne in 1 or
more whorls on the stalk. An unpalatable
grass except for the new growth.

Usually found in slightly damp, sandy
places in Namaqualand, extending into
the Western Cape.
Flowers July to November.

Stipa capensis
steekgras
POACEAE

A sparse, erect annual grass up to 50cm high. The linear leaves are often inrolled. Flowers with long hairs that penetrate animal skin and human clothing are borne close together at the end of the stalk. Usually found in disturbed veld, especially near Vanrhynsdorp and on 'heuweltjies' in Namaqualand, also in other dry winter-rainfall regions of South Africa as well as in northern Africa and the Middle East. **Flowers August to November.**

RIAAN DE VILLIERS

BROMUS
There are 15 species of *Bromus* in southern Africa of which six occur in Namaqualand.

Bromus pectinatus
hawergras
POACEAE   alien

A variable, erect annual grass up to about 50cm high. The leaves are hairy. Long, drooping, hairy flowers are borne sparsely at the top third of a thin erect stalk.

Originating from Asia, it is widespread in Namaqualand, usually in disturbed places, extending to the rest of the winter-rainfall area in the Western and Eastern Cape. **Flowers July to November.**

ZELDA WAHL

STIPAGROSTIS
There are about 30 species of *Stipagrostis* in southern Africa of which 10 occur in Namaqualand.

Stipagrostis ciliata
langbeenboesmangras, growwe twa

POACEAE

A densely tufted perennial grass, with erect stalks up to 1m high, although usually shorter, with a conspicuous ring of stiff white hairs on the nodes of the stalks. The basal, linear leaves are usually curved, rolled and hairless, ending in a narrow point. Flowers are scattered along the top half of the stalk.

Found on sandy flats in Namaqualand, but more common in Bushmanland and other parts of the Karoo and Namibia. **Flowers August to October and February to June.**

NOEL VAN ROOYEN

Stipagrostis obtusa
kortbeenboesmangras, fyn twa, balhaargras

POACEAE

A compact, tufted perennial grass with erect stalks usually up to 50cm high, with a few darkly coloured nodes. The basal, linear leaves are rolled and rigid but recurved. Flowers are borne on the top third of the stalk. A palatable grass.

Found on sandy flats in Namaqualand, but more often in Bushmanland and other parts of the Karoo and Namibia. **Flowers July to May.**

NOEL VAN ROOYEN

Stipagrostis subacaulis
(previously misidentified as *Stipagrostis schaeferi*)
wollerigeblaarboesmangras
POACEAE

A densely tufted perennial grass up to 50cm high, with smooth stalks. The basal and young leaf sheaths are usually woolly.

Found in sandy, gravelly soil in hollows in rocky outcrops, or on gravel plains and along dry watercourses, in the Richtersveld and Namibia.
Flowers August to November and March to June.

NORBERT JÜRGENS

SCHMIDTIA
There are two species of *Schmidtia* in southern Africa of which one occurs in Namaqualand.

Schmidtia kalahariensis
suurgras
POACEAE

A tufted annual grass with erect stalks up to 20cm high. The leaves are expanded at the base while often rolled towards the tips. Flowerheads are packed together at the ends of the stalks. This grass is covered with glandular hairs, making it sticky to the touch and also aromatic, smelling strongly of urine. An unpalatable grass, except when young.

Found in sandy soil in Namaqualand, Bushmanland, the Kalahari and Namibia.
Flowers January to December.

ZELDA WAHL

NOEL VAN ROOYEN

There are two species of *Fingerhuthia* in southern Africa of which one occurs in Namaqualand.

Fingerhuthia africana
vingerhoedgras, borseltjiegras
POACEAE

A densely tufted perennial grass with erect, unbranched stalks usually up to 75cm high. The leaves are expanded at the base, narrowing to a long pointed tip. Flowers are very densely packed for about 2cm at the top of the stalk. An unpalatable grass, but grazed when no other food is available in the veld.

Usually found in rocky places in Namaqualand and also in other areas of South Africa, Botswana and Namibia. **Flowers September to December.**

ANNELISE LE ROUX

PENTAMERIS
There are about 75 species of *Pentameris* in southern Africa of which nine occur in Namaqualand.

Pentameris airoides
POACEAE

A delicate annual grass up to 20cm high. The linear leaves have long glandular hairs. Flowers are widely and sparsely distributed on the stalk.

Found in light shade on lower slopes in Namaqualand, extending through the Karoo to the Drakensberg. **Flowers August to October.**

CLADORAPHIS
There are two species of *Cladoraphis* in southern
Africa and both occur in Namaqualand.

Cladoraphis cyperoides
seebiesie, steekriet

POACEAE

A spiny, tufted perennial grass up to
60cm high. The cylindrical leaves end
in spines. Clusters of flowers are spread
along the stalk. Unpalatable except for
the new growth. It is a good sand-binder,
stabilising wind-blown sea dunes.

Found mostly on white to grey sand
along the coastal dunes of Namaqualand,
extending southwards to the Cape
Peninsula and northwards into Angola.
Flowers August to May.

ZELDA WAHL

Cladoraphis spinosa
volstruisgras

POACEAE

A spiny, woody, shrub-like perennial grass
up to 50cm high. The sword-shaped leaves
are rolled, rigid and sharply pointed.
Flowers are borne in loose clusters along a
spiny stalk.

Found in deep, red sandy flats in
Namaqualand, extending southwards to
Agulhas and northwards into Namibia.
Flowers August to May.

ANNELISE LE ROUX

POACEAE **149**

CHAETOBROMUS
There is one species with three subspecies of
Chaetobromus in southern Africa of which two
subspecies occur in Namaqualand.

Chaetobromus involucratus subsp. *dregeanus*
hartbeesgras

POACEAE

A tufted perennial grass up to 60cm
high. The linear to sword-shaped leaves
are hairless or sometimes sparsely hairy.
Flowers are borne widely spaced at the
top of the stalk. A palatable grass.

Found in sandy areas in Namaqualand,
extending southwards to the Cape Peninsula
and northwards into southern Namibia.
Flowers August to November.

SCHISMUS
There are four species of *Schismus* in southern Africa
and all occur in Namaqualand.

Schismus schismoides
(formerly *Karroochloa schismoides*)
haasgras

POACEAE

A tufted annual grass, spreading flat on the
ground or growing erect up to 15cm high.
The leaf sheath is usually hairless, the linear
leaves are open or rolled, with or without
hairs. Flowers are borne at the top of the
stalk, spreading apart as they mature.

Very common in Namaqualand and
also found in Namibia.
Flowers July to October.

Pennisetum macrourum
jaagbesem, beddinggras
POACEAE

A tufted perennial grass up to 1.5m high,
with creeping rhizomes. The basal leaves,
up to 1m long and 1.5cm across, are pale to
dark green below, with ribbing pronounced
on the upper surface; the margin is
occasionally bluish-purple. Pale green or
straw-coloured flowers, often tinged purple,
are densely packed in a cylinder 12–25cm
long at the top of the stalk.

Found near streams or in damp areas
in Namaqualand and also in the rest of
South Africa and Lesotho.
Flowers November to May.

RIAAN DE VILLIERS

Pennisetum setaceum
fountain grass, klipgras
POACEAE   alien

A tufted perennial grass up to 1m high.
The leaves are up to 40cm long and 3.5mm
across; the margin and upper leaf surface
have small teeth. The pink to purple
flowerheads, up to 25cm long, resemble
bottle-brushes. A weed introduced from
North Africa, without grazing potential,
often used as a garden ornamental but
spreading into the natural vegetation.

Found in disturbed areas, such
as roadsides, in Namaqualand and
throughout South Africa.
Flowers November to July.

RIAAN DE VILLIERS

POACEAE **151**

ZELDA WAHL

TRIBOLIUM
There are about 15 species of *Tribolium* in southern
Africa of which seven occur in Namaqualand.

Tribolium utriculosum

POACEAE

An annual grass with numerous flowering
stalks up to about 10cm high. The linear
leaves have scattered, stiff, white hairs.
Flowers are packed together for 1–1.5cm
at the end of the stalk.

Found in sandy soil on stony slopes
in Namaqualand, with outliers in the
Roggeveld Mountains.
Flowers August to October.

RIAAN DE VILLIERS

CYMBOPOGON
There are about 15 species of *Cymbopogon* in
southern Africa of which five occur in Namaqualand.

Cymbopogon marginatus

scented turpentine grass,
motwortelterpentyngras

POACEAE

A tufted perennial grass up to 80cm high.
The linear leaves are up to 35cm long
and 5mm across and emit a turpentine to
lemony scent when crushed. Flowers are
untidily arranged on the stalk.

Found on rocky slopes in the Kamiesberg
and Bokkeveld Mountains, extending into
the Western and Eastern Cape.
Flowers October to May.

There are two exotic species of *Argemone* in southern Africa, one of which occurs in Namaqualand.

Argemone ochroleuca
Mexican poppy, bloudissel

PAPAVERACEAE   alien

A robust annual herb up to 90cm high, with prickly, greyish stems. The bluish-green leaves, up to 15cm long, are lobed, the lobes toothed with hard, yellowish spines. The pale yellow or cream-coloured flowers are up to 5cm in diameter. The oblong fruits have long spines. A weed introduced from Mexico.

Usually found in dry watercourses or on abandoned lands in Namaqualand and in the rest of southern Africa.
Flowers August to October.

ZELDA WAHL

CYSTICAPNOS
There are three species of *Cysticapnos* in southern Africa of which two occur in Namaqualand.

Cysticapnos vesicaria
klapklappie, klapper

FUMARIACEAE

A soft, climbing or trailing annual herb up to 1m long, with tendrils on the upper stems. The bluish-green leaves, up to 8cm long, are lobed into thin segments. The flowers are up to 8mm long and pink with darker tips. Fruits are thin-walled and inflated.

Grows entwined in other shrubs on stony flats or slopes in Namaqualand, extending eastwards to Calvinia and southwards into the Western Cape.
Flowers August to September.

ZELDA WAHL

ANNELISE LE ROUX

LEUCADENDRON
There are 84 species of *Leucadendron* in southern
Africa of which one occurs in Namaqualand.

Leucadendron brunioides
foetid conebush, tolbos
PROTEACEAE

An erect evergreen shrub up to 2m high,
with male and female flowers on separate
plants. The linear to oblong leaves are up
to 2cm long. Male flowerheads are almost
round and about 1.5cm in diameter; the
yellow female flowerheads are about 1cm
in diameter when mature; the flowers
have a foetid odour. The roundish female
cone is up to 2cm in diameter.

Found in deep, loose sand on the
Coastal Plain and also in the mountains of
the Western Cape.
Flowers October to November.

ANNELISE LE ROUX

VEXATORELLA
There are four species of *Vexatorella* in southern
Africa of which one occurs in Namaqualand.

Vexatorella alpina
Kamiesberg vexator
PROTEACEAE   endemic

A dense, erect or spreading evergreen
shrub up to 1m high. The inversely egg-
shaped to elliptic leaves, blue-green with
a red-brown tip, are up to 4cm long and
1.5cm across. The 2–6 sweetly scented,
creamy-pink flowerheads are borne at the
ends of branches; the flowers are densely
hairy; the style is straight, up to 1.5cm
long and dark carmine.

Found on the upper, rocky slopes of
the Kamiesberg.
Flowers September to December.

LEUCOSPERMUM
There are 48 species of *Leucospermum* in southern
Africa of which two occur in Namaqualand.

Leucospermum praemorsum
pincushion, speldekussing

PROTEACEAE

An erect, evergreen shrub up to 2m high.
The oblong to inversely sword-shaped
leaves taper to a distinct leaf stalk; the
leaf tips are truncate, with 3–5 dark
teeth at the end. Orange to reddish
flowerheads up to 7cm in diameter
are borne at the ends of branches; the
yellow-orange styles are 5–6cm long and
straight to slightly incurved.

Found in deep, red sand on the Coastal
Plain and in the Cederberg.
Flowers July to December.

ANNELISE LE ROUX

Leucospermum rodolentum
beesbos, knopbos

PROTEACEAE

An erect to spreading, evergreen shrub
up to 2m high. The elliptic leaves, densely
covered with silvery-grey hairs, are up
to 6cm long and have 3–6 teeth at the
end. Yellow, sweetly scented flowerheads
about 3cm in diameter are borne in
groups at the ends of branches; the styles
are up to 2cm long and straight.

Found on the Coastal Plain, extending
southwards to the Cape Peninsula.
Flowers August to November.

ZELDA WAHL

ANNELISE LE ROUX

TYLECODON
There are 39 species of *Tylecodon* in southern Africa of which 30 occur in Namaqualand.

Tylecodon occultans
CRASSULACEAE   endemic

A deciduous, dwarf succulent up to 1cm high, the round tuberous base with a thick, flaking skin. The fleshy leaves, already shed at flowering, are up to 1.5cm across and more or less circular, with the stalk attached at a point on the underside and the surface covered with thick, glandular hairs. Pale yellow to yellowish-green flowers are borne on a stalk up to 5cm high; the flower tube is up to 1.5cm long and covered with glandular hairs; the recurved lobes are white to pale pink on the inside.

Found in loamy soil among white quartz pebbles on the Knersvlakte. **Flowers February to March.**

ZELDA WAHL

Tylecodon paniculatus
botterboom, t'kabadda
CRASSULACEAE

A deciduous succulent up to 1.5m high, the very fleshy stem with a flaky, yellowish bark. The fleshy, egg-shaped leaves, already shed at flowering, are up to 12cm long and 10cm across, hairy when young but becoming hairless with age. Orange to red flowers are borne on branched, red stalks, opening in succession; the flower tube is up to 1.5cm long and broadens towards the mouth. This plant, although not often eaten by stock, and then mostly in summer while flowering, causes severe stomach cramps ('krimpsiek') and even death.

Found on rocky slopes in Namaqualand, other drier parts of the Western Cape and also in Namibia. **Flowers November to January.**

156 CRASSULACEAE

Tylecodon pearsonii

CRASSULACEAE   endemic

A deciduous succulent up to 15cm high, the fleshy upper parts of the branches covered with old, short, white leaf bases. The fleshy, sharply pointed, cylindrical leaves, already shed at flowering, are up to 5cm long and straight to sickle-shaped. Yellowish-brown flowers covered with glandular hairs are borne on a sparsely branched stalk up to 10cm long; the flower tube, swollen halfway up, is about 1.5cm long; the lobes are recurved, with reddish striations.

Found in sandy soil, often in quartz, on flats or slopes in Namaqualand. **Flowers November to December.**

ISMAIL EBRAHIM

Tylecodon pygmaeus

CRASSULACEAE   endemic

A deciduous, dwarf succulent up to 8cm high, the erect, fleshy stem with a pale yellow bark. The oblong, fleshy leaves, already shed at flowering, are up to 2cm long and 7mm across and densely covered with large, white, heart-shaped, scale-like hairs that look like crystals. Pale yellowish-green flowers, tinged pink, are borne on a glandular-hairy stalk up to 5cm long; the flower tube is up to 6mm long; the lobes are up to 6mm long.

Found in loamy soil among white quartz pebbles on flats and gentle slopes on the Knersvlakte. **Flowers November to March.**

ADDRIAN FORTUIN

IVAN VAN NIEKERK

Tylecodon racemosus
CRASSULACEAE

A deciduous succulent up to 40cm high, the erect stems simple or branched and covered with old, white, papery leaves that wear away revealing a brown, peeling bark. The fleshy, elliptic to orbicular, flattened leaves, already shed at flowering, are up to 4cm long and blunt to sharply pointed; the leaf surface with glandular hairs or hairless. Flowers are borne on a branched stalk at the ends of branches; they have a pale green flower tube up to 1cm long and white to pale pink lobes up to 6mm long.

Found on stony slopes, in rock crevices and sheltered ravines in Namaqualand, extending northwards into Namibia. **Flowers September to October.**

ANNELISE LE ROUX

Tylecodon reticulatus
sifkopkandelaar
CRASSULACEAE

A deciduous succulent up to 30cm high, the fleshy stems with a yellowish-brown, peeling bark. The fleshy, cylindrical leaves, already shed at flowering, are up to 5cm long and 1cm in diameter, the tips bending inwards. The yellowish-green to brown flower tubes have dark ridges between the fused petals and are up to 8mm long; the yellow to pink, recurved lobes are up to 5mm long; the calyxes and flower stalks dry out after flowering and remain on the plant, forming a persistent, dense, prickly lattice. This plant also causes 'krimpsiek' (see *Cotyledon orbiculata*).

Found on rocky flats and slopes in Namaqualand, Bushmanland, the Western and Eastern Cape and Namibia. **Flowers October to January.**

Tylecodon tenuis
CRASSULACEAE   endemic

A deciduous, straggling shrub up to 15cm
high, with a tuber. The single, erect,
zigzagging stem is green when young and
densely covered with recurved hairs, the
older stem with longitudinal ridges. The
elliptic to inversely egg-shaped leaves,
already shed at flowering, are 6–18mm long
and 4–8mm across and grooved above; the
surface is densely covered with oblong or
club-shaped hairs. The flower tubes, pale
green with thickish red stripes, are 6–8mm
long, 4mm wide at the base, narrowing to
3mm and then expanding to 4mm at the
throat; the lobes are pale green with a fine
network of red stripes; the flower is covered
with club-shaped glandular hairs.

   Found in loamy soil, among white
quartz pebbles, on flats and gentle slopes
on the Knersvlakte.
Flowers November to February.

ISMAIL EBRAHIM

Tylecodon wallichii
krimpsiek, gifkopkandelaar, t'nomsganna
CRASSULACEAE

A deciduous succulent up to 60cm high.
The many erect, dark grey, fleshy stems are
covered with elongated projections that
are the remaining parts of old leaf bases
from previous seasons. The grey-green,
cylindrical leaves are up to 12cm long and
2–4mm in diameter. The drooping flowers
are green to yellowish-green; the flower
tube is 7–12mm long, broadening towards
the mouth; the lobes are 5–7mm long;
the flower is covered with club-shaped
glandular hairs. This plant also causes
'krimpsiek' (see *Cotyledon orbiculata*).

   Found on rocky slopes in
Namaqualand, Bushmanland, the Western
Cape and Namibia.
Flowers December to January.

ZELDA WAHL

IVAN VAN NIEKERK

Cotyledon cuneata
CRASSULACEAE

A sparsely branched, evergreen succulent up to 50cm high. The fleshy, yellow-green leaves, up to 16cm long and 5cm across, are broadly inversely egg-shaped and flattened. Nodding, yellow flowers, glandular-hairy and sticky outside, are borne in a cluster at the top of a stout, sticky stalk; the flower tube, up to 1cm long, is usually shorter than the sharply pointed reflexed or straight lobes, which are up to 2cm long.

Found on rocky slopes in the Richtersveld and Namakwaland Klipkoppe, extending to Victoria West, the Little Karoo and Baviaanskloof.
Flowers November to January.

ZELDA WAHL

Cotyledon orbiculata
pêpê-bos, beesbulk
CRASSULACEAE

An evergreen succulent up to 75cm high. The fleshy, grey-green leaves, up to 10cm long, are oblong to inversely egg-shaped; the margin is thickened and red at the tip. Nodding, orange-red flowers, borne in a loose cluster at the top of a stout stalk, have a flower tube up to 2cm long with recoiling lobes up to 1.2mm long. Hunters use the flower stalk like a flute to make a 'pêpê' sound like that of a young klipspringer. The adults then investigate, making them an easy target. Though not often eaten by stock, and then only in summer, this plant causes severe stomach cramps ('krimpsiek') and even death.

Found on rocky slopes in Namaqualand and other dry areas of the Northern, Western and Eastern Cape and Namibia.
Flowers September to January.

CRASSULA
There are about 160 species of *Crassula* in southern
Africa of which 61 occur in Namaqualand.

Crassula alstonii
bulbees

CRASSULACEAE   endemic

A dwarf succulent up to 2cm high, the
short, rounded stems, usually single but
sometimes in a cluster, completely covered
with flattened, greyish leaves closely
arranged in a spherical rosette. The fleshy
leaves are transversely compressed and
rounded; the leaf surface is covered with
short, thick hairs. Small, cream-coloured to
pale yellow flowers are borne in clusters
at the end of a hairy, branched stalk up to
4cm long.

Found on gentle slopes among white
quartz pebbles in the Richtersveld,
Namakwaland Klipkoppe and on the
Coastal Plain.
Flowers March to April.

ZELDA WAHL

Crassula arborescens
CRASSULACEAE

A much-branched, succulent shrub up
to 1m high, with thick, fleshy stems. The
fleshy, flattened, grey to pinkish leaves,
up to 4cm long and 3cm across, are egg-
shaped to round; the margin is purplish
and slightly thickened. Star-shaped white
to cream-coloured flowers are borne in
a cluster at the top of a stalk up to 7cm
long; the petals are tinged red at the tips
and are fused at the base; the stamens
have purple anthers.

Found on rocky slopes on the
Knersvlakte and in dry areas of the
Western and Eastern Cape.
Flowers October to December.

HEATHER BURGER

IVAN VAN NIEKERK

Crassula barklyi
bandaged finger, seervingerplakkie
CRASSULACEAE

A perennial or biennial dwarf succulent, with simple or branched columnar stems that are entirely covered by tightly clasping leaves; the plants can have short, round branches as illustrated here, or the branches can elongate to about 6cm. The fleshy, brownish, broadly egg-shaped, saucer-like leaves do not bulge outwards; the margin is membranous, with a fringe of hairs. Numerous flask-shaped, cream-coloured flowers are borne in a tight cluster embedded among the leaves; the petals are 1cm long and fused along the lower 2–3mm.

Often, but not always, found among white quartz pebbles or on outcrops on gentle slopes in Namaqualand, extending into the Western Cape.
Flowers May to October.

RIAAN DE VILLIERS

Crassula brevifolia
CRASSULACEAE

A spreading shrub up to 50cm high. The fleshy, linear-elliptic leaves, up to 3cm long and 4mm across, are yellow-green with red tips and upper margins. Yellow flowers are borne in clusters at the end of a stalk up to 5cm long; the petals are fused at the base and the anthers are yellow or black.

Found among granite rocks on slopes, often in exposed conditions, throughout Namaqualand, extending southwards into the Cederberg and northwards into southern Namibia.
Flowers March to June.

Crassula capitella
subsp. *thyrsiflora*
CRASSULACEAE

A succulent up to 25cm high. The short
main stem bears several decumbent
branches ending in leaf rosettes up to 8cm
in diameter; each stem elongates into an
erect flowering stalk. The fleshy leaves,
up to 6cm long, are sword- to egg-shaped
and sharply pointed; the margin has teeth
that point backwards. White or cream-
coloured flowers, tinged red, are borne
in clusters along the stalk; the petals are
fused at the base.

Usually sheltered by rocks or larger
plants on flats or gentle slopes among
quartz pebbles in the Namakwaland
Klipkoppe and on the Knersvlakte,
extending into the Northern, Western and
Eastern Cape and KwaZulu-Natal.
Flowers January to May.

ADRIAN FORTUIN

Crassula columnaris
subsp. *prolifera*
sentkannetjie
CRASSULACEAE

An erect perennial or biennial succulent
up to 10cm high, often with short
branches at the base. The fleshy leaves,
more or less tightly clasping, form a
column 1–4cm in diameter. Numerous
white or pale yellow flowers, often tinged
red, are borne in a tight cluster at the end
of a branch; the stamens have yellow to
brown anthers.

Found on gentle slopes or in
depressions, usually with quartz pebbles,
in Namaqualand, extending eastwards
into Bushmanland and northwards into
southern Namibia.
Flowers June to October.

SOPHIA ETZOLD

PIETER VAN WYK

Crassula corallina subsp. *macrorrhiza*
CRASSULACEAE

A tufted, dwarf succulent up to 3cm high. The closely packed, fleshy, inversely egg-shaped leaves are up to 5mm long and 5mm across; the leaf surface is covered with grey, waxy warts. Cream-coloured flowers are borne in clusters at the end of a stalk; the petals are up to 3.5mm long and fused at the base.

Found on sandy flats, usually among rocks, in the Richtersveld and Namakwaland Klipkoppe, extending eastwards into Bushmanland and northwards into southern Namibia. **Flowers October to January.**

HELGA VAN DER MERWE

Crassula cotyledonis
bergplakkie
CRASSULACEAE

A perennial shrub up to 20cm high. The fleshy, grey-green to yellowish-green, inversely sword- to broadly egg-shaped leaves, up to 6cm long and 3cm across, are densely covered with recurved hairs. Tubular, cream-coloured to pale yellow flowers up to 5mm long are borne in dense, spherical clusters on a hairy stalk up to 40cm long.

Found on rocky slopes in Namaqualand and then from the Little Karoo into the Eastern Cape.
Flowers September to January.

Crassula deceptor
CRASSULACEAE

A dwarf succulent up to 8cm high, usually with many branches, the short branches often congested. The widely ovoid, greyish leaves, sharply pointed to round-tipped, lie tightly packed along the stem, forming a short 4-angled column up to 2.5cm in diameter; the leaves are densely covered with papillae on both surfaces. Tubular, cream-coloured flowers, up to 2.5mm long and fading to brown, are borne on branches at the end of a stalk up to 6cm long.

Found on sandy flats or gentle rocky slopes, often among quartz pebbles, in Namaqualand, extending southwards to Elands Bay, eastwards to Kakamas and northwards into southern Namibia. **Flowers December to March.**

ANNELISE LE ROUX

Crassula dejecta
CRASSULACEAE

An erect, branched shrub up to 35cm high, with hairy stems up to 1cm in diameter. The fleshy leaves, up to 1.5cm long and 1.3cm across, are oblong-elliptic; the margin has nearly spherical papillae. White flowers, tinged red, are borne in a cluster at the end of the stalk; the anthers are brown.

Found on large, rocky outcrops in mountains in the Namakwaland Klipkoppe, extending southwards to Genadendal. **Flowers October to February.**

RIAAN DE VILLIERS

TESSA OLIVER

Crassula dichotoma
CRASSULACEAE

An erect, sparsely branched annual herb up to 15cm high, but usually about 5cm high. The fleshy, inversely egg-shaped to elliptic leaves, up to 1cm long and 5mm across, are pointed or round at the tips and brown-green with red-brown marks. The yellow to orange flowers, up to 2cm in diameter, often have markings of deeper colour in or around the throat of the tube.

Found in sandy soil in Namaqua land, extending southwards into the Western Cape.
Flowers September to October.

ANNELISE LE ROUX

Crassula elegans
CRASSULACEAE

A dwarf succulent up to 8cm high. The very fleshy, green to deep red, egg- to sword-shaped leaves are closely packed on the stem. Whitish flowers, up to 2.5mm long and fading to brown, are borne on a hairy, branched stalk up to 6cm long.

Usually found in shade on flats and rocky slopes in Namaqualand, extending northwards into Namibia.
Flowers December to June.

Crassula expansa
CRASSULACEAE

A perennial herb up to 10cm high, with decumbent, brittle, white to translucent branches up to 15cm long. The fleshy, sword-shaped to elliptic leaves are grooved along the middle of the upper surface. White flowers less than 1cm in diameter are borne singly in the leaf axils and are often clustered towards the ends of branches.

Found on sandy flats or rocky slopes in Namaqualand, extending south- and eastwards into the Western and Eastern Cape, East Africa and also to Madagascar. **Flowers September to December.**

ANNELISE LE ROUX

Crassula hirtipes
CRASSULACEAE   endemic

A shrub up to 15cm high, with brittle, spreading stems. The very fleshy, cylindrical, pointed leaves, up to 1.5cm long and 7mm in diameter, are covered with white, bristle-like, recurved hairs. Cream-coloured to yellow flowers are borne in sparse clusters at the end of hairy stalks up to 2.5cm long; the tips of the lobes are slightly recurved; the stamens have brown to black anthers.

Found in rock crevices or under rocks in shade, on mountain slopes and also on the coast, in Namaqualand. **Flowers August to October.**

ZELDA WAHL

IVAN VAN NIEKERK

Crassula macowaniana
CRASSULACEAE

A much-branched shrub up to 75cm high. The very fleshy, linear to sword-shaped, green to brown leaves, up to 6cm long and 1.5cm in diameter, are flattish above but very convex below; the tips of the leaves often become dry and are shed, leaving truncate ends. White flowers, tinged pink, are borne in clusters on stalks up to 5cm long; the stamens have black anthers.

Found among boulders on rocky slopes in mountainous terrain in Namaqualand, extending into southern Namibia. **Flowers October to December.**

ANNELISE LE ROUX

Crassula muscosa var. *muscosa*
lizard's tail, skoenveterbos
CRASSULACEAE

An erect or scrambling shrub up to 15cm high, the internodes usually not visible between the leaves. The small, blue-green to brown, sword- to egg-shaped leaves are closely packed along the stems. Yellow to yellow-green flowers are borne in leaf axils along the length of the stem.

Found on rocky slopes, often under shrubs, in Namaqualand, extending into the Western and Eastern Cape, Free State and Namibia.
Flowers mainly October to February, but can flower throughout the year.

Crassula namaquensis
CRASSULACEAE

A dwarf succulent up to 10cm high, with short, woody branches ending in rosettes. The very fleshy, blue-green leaves, up to 2.5cm long, are broadly to narrowly elliptic to inversely sword-shaped; the leaf surface is densely covered with very coarse, stubby, recurved hairs. White to yellow flowers are borne in small clusters along a hairy stalk up to 10cm long.

Found on gentle, rocky slopes in the Richtersveld and Namakwaland Klipkoppe, extending eastwards into Bushmanland and northwards into southern Namibia. **Flowers September to November.**

GRETEL VAN ROOYEN

Crassula natans
CRASSULACEAE

An annual herb usually 1–5cm high, but higher if free-standing water is plentiful; the entire plant turns red as soon as it becomes water-stressed. The leaves, up to 1cm long and 3mm across, are inversely egg-shaped to narrowly elliptic, the leaf bases clasping the stem. Tiny, white to red flowers are borne in small clusters all along the stem.

Found in seasonally moist depressions or standing water in the Namakwaland Klipkoppe and widespread throughout South Africa. **Flowers May to November.**

ZELDA WAHL

Crassula plegmatoides
CRASSULACEAE

A dwarf succulent up to 10cm high, usually with only one erect main stem that may become decumbent with age. The grey, widely egg-shaped leaves are strongly concave above, clasping around the stem to form a 4-angled column up to 1.5cm in diameter; the leaf surface is covered with coarse, round, white papillae, with coarse hairs along the margin. Cream-coloured flowers, fading to brown with age, are borne in a sparse cluster at the end of a hairy stalk 3–6cm long.

Found on gentle slopes, usually among quartz pebbles, from Kleinzee on the Coastal Plain, extending northwards into southern Namibia.
Flowers March to April.

ZELDA WAHL

Crassula saxifraga
CRASSULACEAE

A perennial herb up to 15cm high, its erect, unbranched stems arising from 1 to several tubers. The slightly fleshy leaves, up to 3cm long and 7cm across, emerge at flowering time and are green but purplish beneath, the first leaves lying flat on the ground; the margin is irregularly crenate. White or pink flowers are borne in a cluster at the end of a stalk up to 15cm long; the petals are fused at the base; the stamens have brown anthers.

Found on south-facing, rocky slopes in the Richtersveld and Namakwaland Klipkoppe, extending into the Western and Eastern Cape.
Flowers April to June.

IVAN VAN NIEKERK

Crassula strigosa
CRASSULACEAE

An annual herb up to 12cm high, usually
with erect branches covered with long,
spreading hairs. The fleshy, green to
yellowish-green, elliptic, usually hairy
leaves are up to 1.5cm long and 7mm
across. Cup-shaped white flowers, about
2mm in diameter and tinged red, are borne
in sparse clusters at the ends of branches.

This species can be a garden weed.

Found in moist, shaded places in
Namaqualand, extending southwards to
Riviersonderend in the Western Cape.
Flowers August to October.

RIAAN DE VILLIERS

Crassula umbella
CRASSULACEAE

A perennial herb up to 10cm high, with
a tuber. The 2 kidney-shaped, opposite
leaves, up to 7cm long and 10cm across,
are more or less fused into a disc around
the stem; the margin is scalloped. Star-
shaped white to yellowish-green flowers
are borne in clusters on a loosely branched
stalk up to 6cm long.

Found in the shade of overhanging
rocks, usually on south-facing mountain
slopes, in the Richtersveld, Namakwaland
Klipkoppe and also on other mountains in
the Western Cape.
Flowers August to September.

ZELDA WAHL

ANNELISE LE ROUX

ERODIUM
There are eight alien species of *Erodium* in southern Africa of which three occur in Namaqualand.

Erodium cicutarium
horlosie, muskusgras, draaiboor
GERANIACEAE   alien

A prostrate annual or biennial herb up to 8cm high. The deeply double-lobed leaves are up to 9cm long, the surface covered with scattered, minute, white hairs; the margin is toothed. Pink to white flowers are borne in a loose cluster at the end of a long, leafless stalk. The long appendage on the fruit, shaped like the bill of a stork, splits into 5 spiralled awns when ripe, sending the seeds (which have small feathery parachutes) into the air. A weed introduced from Europe.

Found in moist, disturbed places in Namaqualand and in the rest of South Africa and Namibia.
Flowers throughout the year, but peaks August to October.

PIETER VAN WYK

MONSONIA
There are about 35 species of *Monsonia* in southern Africa of which 10 occur in Namaqualand.

Monsonia ciliata
(formerly *Sarcocaulon ciliatum*)
boesmankers
GERANIACEAE   endemic

A spreading shrub up to 18cm high and 30cm in diameter. The fleshy, chestnut- to grey-brown branches are less than 1cm in diameter; the spines are thin, straight and chestnut-brown. The hairy, inversely egg-shaped to elliptic leaves are lobed. The yellow flowers are up to 3.5cm in diameter; the petals are longer than broad and fringed with hairs at the ends.

Found in sandy soils, sometimes among rocks, in the Richtersveld and on the Coastal Plain.
Flowers March to September.

Monsonia crassicaulis
(formerly *Sarcocaulon crassicaule*)
boesmankers
GERANIACEAE

An erect to spreading shrub up to 40cm high. The fleshy, grey or greyish-yellow branches are more than 1cm thick; the spines are straight to slightly curved. The egg-shaped to inversely egg-shaped leaves are lobed, the surface usually covered with fine hairs; the margin is scalloped or toothed. The white to bright yellow flowers are about 3.5cm in diameter.

Found among rocks in Namaqualand, extending eastwards into Bushmanland, southwestwards into the Western Cape and northwards into southern Namibia. **Flowers March to June.**

TED OLIVER

Monsonia flavescens
(formerly *Sarcocaulon flavescens*)
boesmankers
GERANIACEAE

An erect to spreading shrub up to 40cm high and 50cm in diameter. The fleshy, grey-brown to grey-yellow branches are less than 1cm in diameter; the spines are thin, straight and grey-brown. The lobed, broadly egg-shaped to orbicular leaves are slightly or densely covered with soft hairs, or can occasionally be hairless; the margin is toothed. The pale to bright yellow flowers are about 3.5cm in diameter; the petals are longer than broad and without hairs at the ends.

Found on rocky slopes in the Richtersveld and Namakwaland Klipkoppe, extending northwards into southern Namibia. **Flowers June to September.**

ALTA CHAMBERS

GERANIACEAE **173**

PIETER VAN WYK

Monsonia herrei
(formerly *Sarcocaulon herrei*)
boesmankers
GERANIACEAE   endemic

An erect to spreading shrub up to 25cm high and 40cm in diameter. The fleshy, grey-yellow branches are more than 1cm in diameter; the spines, longer than 1.3cm, are thick, straight, hairy and rust-coloured when young. The deeply segmented leaves are covered with long, silky hairs. The pale yellow or white flowers have scattered hairs; the petals are longer than broad.

Found on rocky slopes, usually among quartz rocks below quartz veins in the Richtersveld.
Flowers May to November.

PIETER VAN WYK

Monsonia multifida
(formerly *Sarcocaulon multifidum*)
boesmankers
GERANIACEAE

A spreading shrub up to 20cm high and 25cm in diameter. The fleshy, horizontal, grey-brown to grey-white branches are 1–2cm in diameter and are generally spineless or have short blunt spines up to 6mm long. The deeply segmented leaves are hairy. The flowers are pale pink or magenta with crimson to dark red markings in the centre, but are rarely plain white.

Found in exposed rocky places in the most northern part of the Coastal Plain, extending slightly northwards into southern Namibia.
Flowers March to December.

Monsonia parvifolia
GERANIACEAE

An aromatic, mat-forming perennial herb up to 20cm high and 1m in diameter. The broadly or angularly egg-shaped leaves are covered with glandular hairs; the margin is sometimes pleated. The flowers are white to bright yellow, or pink with reddish veins on the limbs.

Found in sandy soil on riverbeds, roadsides, flats and rocky mountain slopes in the Richtersveld and the Coastal Plain, extending eastwards into Bushmanland and northwards into Namibia.
Flowers February to October.

TED OLIVER

Monsonia patersonii
(formerly *Sarcocaulon patersonii*)
boesmankers
GERANIACEAE

A spreading shrub up to 50cm high. The fleshy, pale grey to almost golden branches are usually more than 1cm in diameter and have thick, straight spines. The blue-green leaves have a notch at the tip. The rose, pale magenta or purple flowers are up to 3.5cm in diameter.

Found in sand, often at the foot of mountains or hills where the sand overlies rocks, in the most northern part of the Coastal Plain, extending northwards into Namibia.
Flowers January to September.

ZELDA WAHL

Monsonia salmoniflora
(formerly *Sarcocaulon salmoniflorum*)
t'noena-doring
GERANIACEAE

An erect to spreading shrub up to 40cm high and 60cm in diameter. The fleshy, pale olive to grey branches are less than 5mm in diameter and have curved or straight spines. The narrowly to broadly elliptic leaves are notched at the tip and hairless. The orange to salmon flowers are about 3cm in diameter; the petals are twice as long as broad.

Found on rocky flats and slopes in Namaqualand, extending into the rest of the Northern and Western Cape and into Namibia.
Flowers September to May.

ZELDA WAHL

Monsonia spinosa
(formerly *Sarcocaulon l'heritieri*)
boesmankers
GERANIACEAE   endemic

A spreading shrub up to 60cm high and 50cm in diameter. The fleshy, pale olive to grey branches, often with vertical stretch markings, are usually less than 1cm in diameter and have long, straight, dark brown to almost black spines. The inversely egg-shaped to round, blue-green leaves have a notch at the tip. The bright yellow flowers are up to 3.5cm in diameter.

Found on rocky slopes, generally in cracks in the rocks or at the base of large granite domes, in Namaqualand.
Flowers August to September.

ZELDA WAHL

PELARGONIUM
There are about 245 species of *Pelargonium* in
southern Africa of which 60 occur in Namaqualand.

Pelargonium barklyi

GERANIACEAE   endemic

A deciduous perennial herb up to 50cm
high, with a tuber. The heart-shaped
leaves lie flat on the ground and have
a purple line near the margin; the veins
on the upper surface are conspicuously
channelled; the underside is purple; the
margin is toothed to lobed. White to
cream-coloured flowers are borne in a
cluster at the end of a stalk up to 50cm
long; the 2 upper petals are slightly
broader than the 3 lower ones.

Found on south-facing slopes, in light
shade of bushes or large rocks in the
Richtersveld and Namakwaland Klipkoppe.
Flowers August to October.

IVAN VAN NIEKERK

Pelargonium carnosum

GERANIACEAE

An erect to spreading shrub up to 75cm
high. The thick, fleshy, yellow-green
branches have a knobbly appearance due
to the slightly swollen nodes. The slightly
fleshy leaves are deeply segmented and up
to 20cm long. The relatively small, white
to greenish-yellow flowers, up to 1.5cm in
diameter, have reddish or purple markings
on the 2 slightly larger upper petals.

Found in sandy soil on flats or slopes in
Namaqualand, extending southwards to the
Little Karoo and northwards into Namibia.
Flowers September to April.

VERONICA ESTERHUIZEN

GERANIACEAE **177**

ANNELISE LE ROUX

Pelargonium crithmifolium
dikbasmalva
GERANIACEAE

A much-branched shrub up to 75cm high. The thick, fleshy, knobbly branches are yellowish-green with a smooth or peeling bark. The fleshy leaves are up to 12cm long, deeply lobed into narrow segments that are again lobed at their ends; the stalks are relatively long and channelled on the upper side. White flowers, about 2cm in diameter and with red markings at the base of the upper petals, are borne in clusters of 4–6; after flowering, the flower stalks dry out and remain as a grey, thorny matrix on the shrub. A palatable to very palatable plant to stock.

Found on flats or rocky slopes in Namaqualand, other mountains in the dry parts of the Northern and Western Cape and in southern Namibia.
Flowers May to October.

HEATHER BURGER

Pelargonium dasyphyllum
GERANIACEAE

A much-branched, deciduous shrub up to 30cm high. The fleshy, smooth, grey-green to greyish-brown branches are up to 1cm in diameter. The fleshy leaves are deeply lobed into narrow segments that are again lobed at their ends; the leaf surface is covered with short glandular hairs. White flowers, with purplish-pink lines on the 2 upper petals, are borne in a loose cluster at the end of a long stalk.

Found on rocky slopes in the Richtersveld, Namakwaland Klipkoppe and also in the Cederberg Mountains.
Flowers August to December.

Pelargonium echinatum
bobbejaan-t'neitjie
GERANIACEAE

A shrub up to 60cm high, with a tuber. The fleshy branches are armed with persistent, recurved, spine-like stipules. The hairy, heart-shaped leaves are up to 3cm long; the margin is scalloped. White, pink or brilliant purple flowers, about 2.5cm in diameter and with darker markings on the 2 upper petals, are borne in a loose cluster at the end of a stalk. A palatable to very palatable plant to stock.

Found on stony slopes in Namaqualand, extending southwards to Clanwilliam in the Western Cape.
Flowers July to November.

Pelargonium fulgidum
rooimalva
GERANIACEAE

A shrub up to 50cm high. The slightly fleshy branches become woody with age. The oblong leaves, up to 10cm long, are lobed and deeply incised; the leaf surface is covered with silvery, silky hairs. Red flowers about 2cm in diameter are borne in a loose cluster at the end of a hairy stalk. A palatable to very palatable plant to stock.

Found on sandy flats or rocky slopes in Namaqualand, extending southwards to Yzerfontein in the Western Cape.
Flowers June to November.

Pelargonium incrassatum
t'neitjie

GERANIACEAE   endemic

A deciduous perennial herb up to 30cm high, with a tuber. The soft leaves, in a basal rosette, are up to 14cm long and lobed into scalloped segments; the leaf surface is covered with silvery hairs. Brilliant pinkish-purple flowers about 2cm in diameter are borne in a cluster at the end of a stalk; the 2 upper petals are much larger than the 3 lower ones. The local inhabitants consider the tuber a delicacy.

Found on rocky flats or slopes in the Namakwaland Klipkoppe.
Flowers August to September.

Pelargonium klinghardtense
GERANIACEAE

A spreading deciduous shrub up to 60cm high. The fleshy branches are smooth, blue-green to pale yellowish-green, and there are 2 kinds: long branches and leaf-bearing dwarf shoots. The fleshy, faintly aromatic, pale blue-green leaves are undulate. White flowers are borne on a sparsely branched stalk that hardens after flowering, persisting for some time.

Found on rocky outcrops, often wedged among rocks, in the Richtersveld, extending into southern Namibia.
Flowers May to September.

Pelargonium nanum
GERANIACEAE

A prostrate annual herb with radiating branches up to 50cm long. The egg- to heart-shaped leaves are up to 2cm across; the margin is deeply as well as shallowly lobed. The flowers are about 8mm in diameter; the 2 upper petals are white with dark wine-red markings and the 3 lower ones are pink.

Found on rocky slopes in the Namakwaland Klipkoppe, extending into the Western Cape.
Flowers late August to December.

ANNELISE LE ROUX

Pelargonium praemorsum
GERANIACEAE   endemic

A spreading shrub up to 50cm high. The shiny brown, zigzag branches are sparsely covered with very short hairs and have short, hardened, thorn-like stipules. The kidney-shaped leaves, up to 2cm long, are deeply divided into 5 wedge-shaped, 3-toothed segments with minute hairs. Cream-coloured flowers, up to 4cm in diameter and with reddish-purple feather-like markings on the 2 larger upper petals, are borne singly or in pairs at the ends of branches; the stamens are sharply recurved. A palatable to very palatable plant to stock.

Found on rocky slopes in Namaqualand.
Flowers August to April.

ANNELISE LE ROUX

ZELDA WAHL

Pelargonium pulchellum
GERANIACEAE   endemic

A deciduous perennial herb up to 50cm high, with short, fleshy branches. The oblong to egg-shaped leaves, up to 15cm long, are irregularly lobed and incised; the leaf surface is covered with hairs. White flowers, about 2.5cm in diameter and with red markings on the 3 lower petals, are borne in a cluster at the end of a stalk. A palatable to very palatable plant to stock.

Found on rocky slopes in the Namakwaland Klipkoppe.
Flowers July to October.

ANNELISE LE ROUX

Pelargonium quarciticola
GERANIACEAE   endemic

An erect, deciduous perennial herb up to 10cm high, with a tuber. The slightly fleshy, erect leaves are deeply and finely incised; the leaf surface is covered with sticky glandular hairs. White or pale pink flowers, with red feather-like markings on the 2 larger upper petals, are borne 2 or 3 together at the end of a stalk up to 10cm long.

Found on gentle slopes among white quartz pebbles on the Knersvlakte.
Flowers July to August.

Pelargonium rapaceum
GERANIACEAE

A deciduous perennial herb up to 30cm high, with a tuber. The long, erect, linear and very hairy leaves, already shed at flowering, are much dissected and incised. Flowers are borne in a cluster at the end of a stalk; they are cream-coloured, yellow, primrose-yellow or pink with red stripes on the lower part of the upper petals; the 2 upper petals are bent back and the 3 lower ones are close together, enclosing the anthers completely. In the past, San people ate the tubers, which they roasted in hot ashes, and it is still used as an antidiarrhoeic today.

Found in sandy soil and often on rocky mountain slopes in Namaqualand, extending into the Western Cape. **Flowers October to February.**

ANNELISE LE ROUX

Pelargonium scabrum
hoenderbos
GERANIACEAE

An erect shrub up to 1.5m high, with rough branches. The leaves are hard, rough to the touch, about 4cm long, shallowly divided into toothed lobes and lemon-scented when bruised. White to mauve flowers, with purple feather-like markings on the 2 larger upper petals, are borne in clusters of 2–6. The common name alludes to the shape of the leaves, which resemble chicken feet. A palatable to very palatable plant to stock.

Found on rocky slopes in the Namakwaland Klipkoppe and other mountainous areas of the Western and Eastern Cape.
Flowers August to January.

ZELDA WAHL

Pelargonium sericifolium
naaldekoker

GERANIACEAE   endemic

A much-branched, evergreen shrub up to 20cm high. The wedge-shaped leaves are up to 1cm long and lobed at the end; the leaf surface is covered with dense, silvery hairs. Stalked cerise flowers, about 1.5cm in diameter and with a darker band on the 3 smaller lower petals, are borne in clusters of 1–3. A palatable to very palatable plant to stock.

Found in shale on rocky slopes in the Namakwaland Klipkoppe.
Flowers July to October.

Pelargonium spinosum
gifdoring, noerap

GERANIACEAE

An erect, deciduous shrub up to 1m high, with yellowish, peeling bark. The slightly fleshy, aromatic, pale green to blue-green leaves, up to 2.5cm long and across, are borne on stalks up to 10cm long that remain on the plant after the leaves have dropped, turning into spines; the margin is toothed. White to pale pink flowers, about 2cm in diameter and with purple feather-like markings on the 2 upper petals, are borne in clusters of 3–10.

Found on rocky slopes in the Richtersveld and Namakwaland Klipkoppe, extending eastwards into Bushmanland and northwards into southern Namibia.
Flowers July to September.

Pelargonium triste
rasmusbas
GERANIACEAE

A deciduous perennial herb up to 50cm high, with a tuber. The basal, flat-lying leaves are up to 40cm long and finely dissected into numerous hairy segments (like carrot leaves). Cream-coloured, yellowish-green or purplish-brown flowers about 1.7cm in diameter are borne in a sparse cluster at the end of a stalk 40cm long; the flowers are scented at night. The red colour in the tuber is used to dye leather. A palatable to very palatable plant to stock.

Found in sandy soil, often on abandoned agricultural fields, in Namaqualand, extending into the Western Cape.
Flowers August to November.

MELIANTHUS
There are eight species of *Melianthus* in southern Africa of which two occur in Namaqualand.

Melianthus pectinatus
kruidjie-roer-my-nie

MELIANTHACEAE   endemic

An erect shrub up to 1.5m high. The leaves, up to 19cm long, are divided into 11–15 leaflets that are linear to sword-shaped, entire to scalloped; the midrib is winged; the leaves emit a very pungent odour when touched. Scarlet flowers about 1.5–2cm long are borne in loose clusters at the ends of branches. Fruits are 4-winged and hairless.

Found on rocky slopes or in dry riverbeds in the Richtersveld and the Namakwaland Klipkoppe.
Flowers May to September.

Melianthus elongatus found on the Coastal Plain of Namaqualand has a more deeply crenate leaf margin and densely hairy fruits.

ZELDA WAHL

Roepera cordifolia
(formerly *Zygophyllum cordifolium*)
sjielingbos, geldjiesbos
ZYGOPHYLLACEAE

A spreading shrub up to 35cm high. The slightly fleshy, blue-green, opposite leaves, up to 3.5cm long, are round to egg-shaped or inversely egg-shaped and stalkless. The yellow flowers with red or brown marks in the centre are about 3cm in diameter and have 5 petals. The drooping fruits are oblong, with 5 prominent wings.

Found in sandy soil on the Coastal Plain and Knersvlakte, extending southwards to Saldanha and northwards into Namibia. **Flowers April to January.**

ZELDA WAHL

Roepera foetida
(formerly *Zygophyllum foetidum*)
skilpadbos
ZYGOPHYLLACEAE

A sprawling shrub up to 1.5m high. Branches are square in cross-section. The bright green leaves, borne on a stalk up to 1cm long, are divided into 2 slightly fleshy leaflets that are inversely egg-shaped with an obliquely narrow base. The yellow flowers with red or brown marks in the centre are about 3cm in diameter and have 5 petals. The drooping fruits are oblong and 5-angled. This plant exudes a pungent odour in the veld.

Found on rocky slopes in Namaqualand, extending southwards into the Western and Eastern Cape.
Flowers August to December.

Roepera leptopetala
(formerly *Zygophyllum leptopetalum*)
ZYGOPHYLLACEAE

A much-branched shrub up to 1m high. Branches are round in cross-section; the young reddish branches have longitudinal ridges. The blue-green leaves, borne on a stalk up to 1.1cm long, are divided into 2 slightly fleshy leaflets that are inversely egg-shaped with an oblique base. The white flowers with reddish-maroon veins and blotches in the centre are about 2cm in diameter and have 5 petals. The drooping fruits are round to slightly oblong. A very palatable plant to stock.

Found in Namaqualand, extending into southern Namibia.
Flowers July to October.

ANNELISE LE ROUX

Roepera morgsana
(formerly *Zygophyllum morgsana*)
skilpadbos
ZYGOPHYLLACEAE

An erect, much-branched shrub up to 1.5m high. The blue-green leaves, borne on a stalk up to 3mm long, are divided into 2 slightly fleshy, inversely egg-shaped leaflets. Yellow flowers with red or brown markings in the centre have 4 petals and are borne in pairs at the ends of branches. The drooping fruits have 4 prominent membrane-like wings. An unpalatable plant to stock, except for the new growth, which usually appears earlier than on other shrubs.

Found on sandy flats and rocky slopes in Namaqualand and also in the dry parts of the Western Cape and in Namibia.
Flowers May to November.

ZELDA WAHL

ZELDA WAHL

Roepera sp.
(formerly *Zygophyllum* sp.)
dwergskilpadbos

ZYGOPHYLLACEAE   endemic

An erect to spreading shrub up to 60cm
high. The older, dark grey branches are
cracked and twisted; the nodes are slightly
swollen. The green or blue-green, stalkless
leaves are divided into 2 slightly fleshy
leaflets that are inversely egg-shaped,
5–7mm long and 2–3.5mm across. The
pale yellow flowers with reddish-brown to
brown-purple markings on the inside are
about 1cm in diameter and have 5 petals.
The drooping fruits are round to slightly
oblong, 5-ridged and with a point at the
end. This species is well known but still has
to be formally named and described. A
palatable plant to stock.
   Found in sandy soil in the Richtersveld
and Namakwaland Klipkoppe.
Flowers July to September.

ANNELISE LE ROUX

Roepera teretifolia
(formerly *Zygophyllum teretifolium*)

ZYGOPHYLLACEAE   endemic

A prostrate to spreading shrub up to 40cm
high. The very fleshy, cylindrical leaves
are green becoming a burgundy colour
when the dry season starts; the leaves are
divided into 2 leaflets that are rounded
at the tips, up to 2.5cm long and 4mm
in diameter. The nodding, pale yellow
flowers are marked with deeper yellow,
brown or red markings in the centre and
have 5 petals. The drooping, oblong,
burgundy-coloured fruits have 5 ridges.
   Found among white quartz pebbles on
the Knersvlakte.
Flowers June to July.

AUGEA
There is only one species of *Augea* in southern Africa
and in Namaqualand.

Augea capensis
boesmandruiwe, kinderpieletjies

ZYGOPHYLLACEAE

An erect, short-lived shrub up to 50cm
high. The yellow branches are fleshy. The
very fleshy, yellow-green leaves, up to
4cm long, are cylindrical. Whitish-green
flowers are borne singly on the nodes of
the branches. The fleshy, cylindrical fruits,
sometimes mistaken for leaves in the
green immature stages, are about 2cm
long and turn yellow when ripe.

Usually found in disturbed places in
Namaqualand, Bushmanland, the Tankwa
Karoo, Little Karoo and Namibia.
Flowers January to December after rain.

ANNELISE LE ROUX

TETRAENA
There are 15 species of *Tetraena* in southern Africa of
which about 10 occur in Namaqualand.

Tetraena clavata
(formerly *Zygophyllum clavatum*)

ZYGOPHYLLACEAE

A spreading shrub up to 20cm high.
The small, very fleshy leaves are divided
into 2 leaflets that are club-shaped and
asymmetrical. The small, white or yellowish
flowers have 5 petals. The round, fleshy
fruits, divided into 5 segments, are strongly
horizontally compressed.

Found in sandy soil or rocky flats near
the coast on the Coastal Plain, extending
from Port Nolloth into southern Namibia.
Flowers January to December after rain.

NORBERT JÜRGENS

Tetraena microcarpa
(formerly *Zygophyllum microcarpum*)
ZYGOPHYLLACEAE

An erect to spreading to prostrate shrub up to 30cm high, with minute white, 2-armed hairs on new growth. Branches are yellowish, grey below, and thickened at the joints. The bright green leaves, borne on a stalk up to 8mm long, are divided into 2 slightly fleshy leaflets that are narrowly elliptic. White or pale yellow flowers with 5 spoon-shaped petals are borne in leaf axils. The softly hairy fruits are broadly 5-winged, deeply incised at the top; the wings are twisted like a propeller. This plant is poisonous to stock.

Found in sandy soil in the Richtersveld, central Karoo, Little Karoo and in Namibia. **Flowers January to December after rain.**

ANNELISE LE ROUX

Tetraena prismatocarpa
(formerly *Zygophyllum prismatocarpum*)
penbos
ZYGOPHYLLACEAE

An erect, sparsely branched shrub up to 1.5m high. The young branches are longitudinally 1-winged. The grey-green to yellowish, opposite, slightly fleshy and leathery leaves are stalkless and inversely egg-shaped, narrowing to the base. White flowers are borne at the ends of branches; the petals are only slightly longer than the calyx; the stamens are longer than the petals. The erect to drooping fruits are oblong and strongly 5-angled.

Found in dry watercourses, on rocky flats and gentle slopes in the Richtersveld and on the Coastal Plain north of Kleinzee, extending into southern Namibia. **Flowers July to October.**

ANNELISE LE ROUX

Tetraena retrofracta
(formerly *Zygophyllum retrofractum*)
hondepisbos
ZYGOPHYLLACEAE

A rounded shrub up to 50cm high.
Branches and twigs arch backwards. The
blue-green leaves, borne on short, fleshy
stalks, are divided into 2 fleshy, inversely
egg-shaped leaflets up to 5mm long.
The small flowers are white; the petals
are only slightly longer than the calyx;
the stamens are longer than the petals.
Fruits are oblong to spherical and slightly
5-angled.

Found in sandy soil along riverbanks,
on flats or slopes, often in disturbed areas,
in Namaqualand, extending southwards
to the Ceres Karoo and northwards into
southern Namibia.
Flowers July to December.

PIETER VAN WYK

Tetraena simplex
(formerly *Zygophyllum simplex*)
volstruisdruiwe
ZYGOPHYLLACEAE

A prostrate, mat-forming annual or
biennial herb up to 1m in diameter. The
green, yellowish or sometimes maroon,
fleshy leaves are cylindrical to roundish.
The small, yellowish flowers have 5 petals.
The fleshy, 5-parted fruits break up into 5
sections when dry.

Found in a variety of soils mostly
on flats, often in disturbed areas, in
Namaqualand, extending eastwards into
Botswana and northwards into Namibia.
Found outside southern Africa in Egypt
and Arabia.
Flowers January to December after rain.

NORBERT JÜRGENS

ZYGOPHYLLACEAE **191**

SISYNDITE

There is only one species of *Sisyndite* in southern Africa and in Namaqualand.

Sisyndite spartea
desert broom, t'kambos
ZYGOPHYLLACEAE

An erect, deciduous, sparsely and thinly branched, blue-green shrub up to 1.5m high. The blue-green leaves have very thick midribs that look like spiny branches; the small, oblong leaflets, widely spaced on the midrib, are very short-lived. Yellow flowers are borne singly in leaf axils. The 5-angled fruits have thick, long, silky, cream-coloured to yellow hairs.

Found in sandy soil, often in dry riverbeds and on roadsides, in the Richtersveld, extending eastwards into Bushmanland and northwards into Namibia. **Flowers January to December after rain.**

TRIBULUS

There are five species of *Tribulus* in southern Africa of which four occur in Namaqualand.

Tribulus pterophorus
duifiedoring
ZYGOPHYLLACEAE

A prostrate annual herb with a rootstock, all vegetative parts more or less densely hairy. The leaves of varying sizes are divided into 6–9 pairs of oblong leaflets in the larger leaves and only 3–6 pairs in the smaller leaves; leaflets are densely silky beneath, less so on the upper surface. Flowers bright yellow to orange. Fruits up to 1cm in diameter have 5 large wings with several small, short, thick spines in between the wings.

Found in sandy soil in the Richtersveld, extending eastwards into the rest of the Northern Cape and northwards into Namibia. **Flowers September to May.**

192 ZYGOPHYLLACEAE

Tribulus terrestris
duwweltjie
ZYGOPHYLLACEAE

A prostrate annual or perennial herb with trailing branches up to 75cm long; all vegetative parts have long or bristly hairs, but vary in the degree of hairiness. The leaves of varying sizes are divided into 12 pairs of oblong leaflets in the larger leaves and only up to 8 pairs in the smaller leaves, the pairs of leaflets not borne exactly opposite each other and covered with long hairs. Yellow flowers, up to 2cm in diameter but often smaller, are borne on short stalks. Fruits are usually hairy; each of the 5 segments has 2 lateral, spreading spines near the top and shorter spines directed downwards.

Found in sandy soil on flats, often in disturbed sites, in Namaqualand and also widespread in the rest of southern Africa. **Flowers October to April.**

ANNELISE LE ROUX

Tribulus zeyheri
duwweltjie
ZYGOPHYLLACEAE

A prostrate annual or perennial herb with trailing branches up to 25cm long; all vegetative parts are more or less hairy with scattered, bristle-like, bulbous-based hairs. The leaves, of varying sizes, are divided into 9 pairs of oblong leaflets in the larger leaves and only up to 4 pairs in the smaller leaves; the leaflets have long hairs on the margin. Yellow or white flowers up to 5cm in diameter are borne on stalks up to 5cm long. The fruits have many strong, narrow, conical spines, rarely spineless but warty.

Found in sandy soil on flats in Namaqualand and also widespread in the rest of southern Africa. **Flowers January to May.**

ANNELISE LE ROUX

ZELDA WAHL

ADENOLOBUS
There are two species of *Adenolobus* in southern Africa of which one occurs in Namaqualand.

Adenolobus garipensis
bloubeesklou
FABACEAE

A sparsely branched small tree or large shrub up to 3m tall. The erect branches are long, slender and rod-like. The leaves, up to 2cm long and 3cm across, are shallowly bilobed. Reddish to yellow flowers with reddish veins are borne in clusters of 1–3; the stamens are longer than the petals. The kidney-shaped, flat fruit pods have conspicuously stalked glandular hairs.

Found on sandy flats or rocky slopes, often in dry riverbeds, in the Richtersveld, extending eastwards to the Augrabies Falls and northwards into southern Angola. **Flowers August to April.**

ANNELISE LE ROUX

PROSOPIS
There are five alien species of *Prosopis* in southern Africa of which two occur in Namaqualand.

Prosopis velutina
mesquite, soetdoring, meskietboom
FABACEAE   alien invader

A tree up to 3m tall. The sparsely to densely hairy leaves are divided into a pair of leaflets, each of which is again divided into 13–20 pairs of smaller leaflets. Yellow flowers are closely packed on a long stalk. The cream-coloured to pale yellow, hairy fruit pods are straight or sickle-shaped; the margin is slightly constricted between the seeds. The pods are used for fodder, a practice that spreads the seeds.

Usually found along drainage lines in Namaqualand and also widespread in the rest of southern Africa. Originally from North and Central America. **Flowers June to November.**

Vachellia karroo

(formerly *Acacia karroo*)

soetdoringboom

FABACEAE

A branched tree or large shrub up to 5m tall. Branches bear straight, white thorns 4–10cm long. Leaves are divided into 2–6 pairs of leaflets, each of which is again divided into 5–15 pairs of smaller leaflets; a gland is usually present on the leaf stalk. Small, sweetly scented, yellow flowers are closely packed in a round cluster. The hairless fruit pods are straight or sickle-shaped; the margin is slightly constricted between the seeds.

Usually found in dry riverbeds in Namaqualand and also widespread in the rest of southern Africa.
Flowers October to February.

ANNELISE LE ROUX

Aspalathus angustifolia subsp. *robusta*

FABACEAE   endemic

A rigid, erect to spreading shrub up to 1m high. Young branches are pale yellow. The stiff leaves have 3 prominent veins and end in a sharp point; the margin is transparent. Yellow flowers with dense, white, silky hairs on the outside of the petals are borne singly in leaf axils. The hairless fruit pods are linear, straight and up to 3cm long.

Found on rocky ridges on top of the Kamiesberg.
Flowers September to October.

ANNELISE LE ROUX

There are about 10 species of *Wiborgia* in southern Africa of which six occur in Namaqualand.

Wiborgia monoptera
wolfdoring, ertjiebos
FABACEAE

A much-branched, erect shrub up to 1m high. Young branches are hairy and the branches end in sharp, spine-like tips. The almost hairless or sparsely silvery-hairy leaves are divided into 3 leaflets with sharp tips. The pale yellow flowers are about 1cm long. The round to broadly oval fruit pods are slightly inflated, with a flat rim; the sides are veined. A palatable to very palatable plant to stock.

Found on rocky slopes in Namaqualand, extending southwards to Clanwilliam. **Flowers July to September.**

ZELDA WAHL

Wiborgia obcordata
kinnabos
FABACEAE

An erect, sparse, slender shrub up to 3m tall. Young branches are hairy. The oblong leaves are divided into 3 leaflets. The yellow flowers have a prominent, pointed keel. Fruit pods have a raised upper ridge and the slightly inflated sides have a few prominent veins.

Found in deep, red, sandy soil in Namaqualand, extending into the Western Cape. **Flowers August to March.**

ANNELISE LE ROUX

196 FABACEAE

CALOBOTA
There are 16 species of *Calobota* in southern Africa of
which seven occur in Namaqualand.

Calobota halenbergensis
(formerly *Lebeckia halenbergensis*)
fluitjiesbos

FABACEAE

An erect, lax shrub up to 75cm high. The
yellow-green leaves, borne on a stalk
up to 3cm long, are divided into 3 linear
leaflets that are up to 1.5cm long. The
yellow flowers are up to 10mm long. The
oblong fruit pods are up to 1.5cm long. A
palatable plant to stock, but mainly the
new growth and especially the flowers
and pods are eaten.

Found on sandy flats in Namaqualand, extending northwards into
southern Namibia.
Flowers June to November.

ZELDA WAHL

Calobota sericea
(formerly *Lebeckia sericea*)
bloufluitjiesbos, t'aibie

FABACEAE

A densely branched shrub up to 1.5m
high. Young branches are covered with
silver hairs. The blue-green leaves, borne
on a stalk up to 5.5cm long, are divided
into 3 linear to elliptic leaflets that are up
to 3.5cm long; the surface of the leaflets
is covered with minute silvery hairs. The
bright, deep yellow or cream-coloured
flowers are about 1.5cm long. The young
cylindrical fruit pods are covered with
minute silky hairs. A palatable plant to
stock, but mainly the new growth and
especially the flowers and pods are eaten.

Found on rocky slopes in Namaqualand,
extending southwards to Clanwilliam.
Flowers July to October.

ANNELISE LE ROUX

ANNELISE LE ROUX

LOTONONIS
There are about 90 species of *Lotononis* in southern
Africa of which 15 occur in Namaqualand.

Lotononis densa subsp. *gracilis*
FABACEAE

An erect shrub up to 60cm high. The
leaves, borne on a stalk 2–3 times longer
than the leaflets, are divided into 3
leaflets that are slightly folded; the upper
surface of the leaflets is hairless. Yellow
flowers are sparsely arranged at the ends
of branches. The ovoid to broadly oblong
fruit pods are inflated; the upper suture is
slightly warty.

Found on rocky slopes in the
Namakwaland Klipkoppe.
Flowers June to October.

ZELDA WAHL

Lotononis falcata
FABACEAE

A prostrate annual herb with branches up
to 25cm long, but usually much shorter.
The leaves are divided into 3 oblong,
minutely hairy leaflets that are up to 1cm
long. White, cream-coloured or yellow
flowers are borne singly or in clusters of 3
or more. The hairy fruit pods are oblong
or often slightly sickle-shaped.

Found in sandy soil, often in disturbed
areas, in Namaqualand, extending
eastwards to Sutherland, northwards into
southern Namibia and also found in the
central Karoo.
Flowers June to October.

Lotononis strigillosa
FABACEAE

A prostrate annual herb with hairy branches up to 5cm long. The leaves are divided into 3 broadly inversely egg-shaped leaflets; the surface of the leaflets is covered with evenly spaced, short, thick hairs. The yellow flowers are up to 1cm long. Fruit pods are long, flat and distinctly sickle-shaped.

Found in sandy soil in the Richtersveld, extending into southern Namibia. **Flowers July to October.**

ZELDA WAHL

EUCHLORA
There is one species of *Euchlora* in southern Africa and in Namaqualand.

Euchlora hirsuta
(formerly *Lotononis hirsuta*)
noorsbal
FABACEAE

A perennial herb up to 15cm high with hairy annual branches. The narrowly elliptic leaves, with long sparse hairs, are usually simple and stalkless, at least on the basal parts of the flowering branches. Yellow flowers are borne close together at the ends of branches. The oblong fruit pods are strongly inflated laterally; the upper suture is slightly warty.

Found in clayey soil in the Namakwaland Klipkoppe and on other mountains in the Northern and Western Cape. **Flowers July to October.**

TESSA OLIVER

ANNELISE LE ROUX

LEOBORDEA
There are about 40 species of *Leobordea* in southern Africa of which nine occur in Namaqualand.

Leobordea digitata
(formerly *Lotononis digitata*)
FABACEAE

A prostrate, spreading shrub with branches up to 50cm long. The leaves, divided into 5 elliptic leaflets, are minutely and inconspicuously hairy, appearing hairless. The bright yellow flowers are up to 1.2cm long. The flat fruit pods are distinctly sickle-shaped.

Found on flat granite rock sheets in the Namakwaland Klipkoppe and in the Kouga Mountains.
Flowers August to November.

GRETEL VAN ROOYEN

Leobordea longiflora
FABACEAE   endemic

A prostrate, spreading shrub with branches up to 20cm long. The leaves are divided into 5 narrowly oblong leaflets that are folded inwards; the leaflet surface is densely hairy. The large, showy, slender, yellow flowers are more than 2cm long; the sharply pointed keel petal is very large, much longer than the standard and wing petals. Fruit pods are borne on a long stalk.

Found on flat granite rock sheets in the Namakwaland Klipkoppe.
Flowers September to October.

Leobordea polycephala

FABACEAE   endemic

A prostrate annual herb with branches up
to 30cm long. The leaves are divided into
3 elliptic leaflets that are folded slightly
inwards; the leaflet surface is covered
with white, silky hairs. Yellow flowers
more than 2cm long are borne in a round
cluster; the keel petal is much longer than
the standard and wing petals; all parts of
the flower are covered with long, white
hairs. Fruit pods are enclosed by a much-
inflated, very hairy calyx.
    Found in sandy soil on the Kamiesberg.
Flowers September to October.

Leobordea quinata

FABACEAE   endemic

A prostrate, spreading shrub with thin,
rigid branches up to 20cm long. The leaves
are usually divided into 5 elliptic leaflets
that are folded slightly inwards; the leaflet
surface is covered with hairs. The yellow
flowers have minute hairs along the back
midline of the standard petal. The oblong
fruit pods are slightly inflated.
    Found in sandy soil or in cracks of rock
sheets in the Namakwaland Klipkoppe.
Flowers August to November.

ANNELISE LE ROUX

CROTALARIA
There are about 60 species of *Crotalaria* in southern Africa of which five occur in Namaqualand.

Crotalaria excisa subsp. *namaquensis*

FABACEAE   endemic

A prostrate, spreading perennial herb with branches up to 20cm long. The leaves are divided into 3 elliptic leaflets that are folded slightly inwards. Relatively large, yellow flowers are borne on an upright stalk up to 15cm long. The oblong fruit pods are inflated.

Found in sandy soil in flat areas in Namaqualand, extending southwards to Malmesbury.
Flowers June to October.

ZELDA WAHL

Crotalaria meyeriana
sandklawer
FABACEAE

A prostrate, spreading perennial herb with branches up to 25cm long and covered with long, silver hairs. The leaves, borne on a stalk up to 2.5cm long, are divided into 3 inversely egg-shaped leaflets covered with long, spreading hairs on both surfaces. The yellow flowers are up to 1.5cm long; the largest petal has reddish-brown veins, with hairs along the back midrib. The inflated, oblong fruit pods are up to 2cm long and covered with long, spreading hairs.

Found in sandy soil in the Richtersveld, extending northwards into southern Namibia.
Flowers April to October.

MELOLOBIUM
There are about 15 species of *Melolobium* in southern
Africa of which four occur in Namaqualand.

Melolobium adenodes
ertjiebos
FABACEAE

An erect shrub with red-brown branches,
up to 30cm high. The twigs and leaves
are covered with stalked glandular hairs,
giving them a rough appearance. The flat,
sticky leaves are divided into 3 broadly
inversely egg-shaped leaflets. The flowers
are yellow, with stamens fused into a
closed tube. The oblong fruit pods, with
stalked glandular hairs, are slightly sickle-
shaped at the ends and often slightly
constricted between the seeds.

Found in sandy to loamy soil in
Namaqualand, extending southwards
to Piketberg and also found in the
central Karoo.
Flowers August to October.

ZELDA WAHL

Melolobium candicans
ertjiebos
FABACEAE

A spreading, rigid shrub up to 40cm high,
with branches ending in sharp spines and
densely covered with short, white hairs
that give them a white appearance. The
leaves are divided into 3 oblong leaflets
that are folded slightly inwards. Yellow
flowers, darkening to reddish-orange with
age, are borne at the ends of branches.
The oblong fruit pods are densely covered
with straight hairs.

Found on rocky flats or slopes in
Namaqualand, the rest of the Northern
Cape and in the Western and Eastern
Cape, Free State and southern Namibia.
Flowers July to December.

HEATHER BURGER

ANNELISE LE ROUX

There are nine species of *Indigastrum* in southern Africa of which two occur in Namaqualand.

Indigastrum argyroides
FABACEAE

A prostrate, spreading annual herb up to 2cm high, with branches densely covered with short, stiff hairs pressed to the surface. The leaves are divided into 2 or 3 leaflets of which the terminal one is at least twice the size of the lateral ones. The flowers are magenta. Fruit pods are 1.5cm long.

Found in open sandy and rocky flats in the Richtersveld, extending northwards into southern Namibia.
Flowers September to May.

RIAAN DE VILLIERS

INDIGOFERA
There are about 205 species of *Indigofera* in southern Africa of which 14 occur in Namaqualand.

Indigofera amoena
FABACEAE

A suberect to spreading woody herb or shrub up to 60cm high. The leaves are divided into 3 elliptic-oblong or inversely egg-shaped leaflets, thinly covered with hairs. Pink or magenta flowers are tightly packed at the ends of stalks up to 40cm long. The oblong fruit pods are covered with short, sparse hairs.

Found in sandy soil in the Namakwa-land Klipkoppe and on the Knersvlakte, extending southwards to Piketberg.
Flowers July to October.

Indigofera auricoma
FABACEAE

A prostrate to spreading annual or perennial herb with branches up to 40cm long and densely covered with short, stiff hairs pressed to the surface. The leaves are divided into 4–11 elliptic leaflets that are folded slightly inwards. Magenta to purplish-red flowers are closely packed at the end of a stalk 5cm long. Fruit pods are oblong.

Found in sandy soil, often in dry watercourses, in the Richtersveld, extending eastwards into Bushmanland and northwards into Namibia.
Flowers July to May.

SUE HOPPER

Indigofera nigromontana
FABACEAE

A much-branched, erect shrub up to 2m high, with the branches ending in spines. The leaves are divided into 3 inversely egg-shaped leaflets, folded slightly inwards and thinly hairy above. The flowers are magenta to pinkish-purple. The hairless, oblong pods are up to 2cm long.

Found on rocky slopes in the Namakwaland Klipkoppe, extending into the Western and Eastern Cape and into Lesotho.
Flowers March to October.

ANNELISE LE ROUX & CAMERON MCMASTER

FABACEAE **205**

LEONARD SEGAL

TEPHROSIA
There are about 30 species of *Tephrosia* in southern Africa of which one occurs in Namaqualand.

Tephrosia dregeana
FABACEAE

An annual herb up to 30cm high, with greyish, minutely hairy branches. The leaves are divided into 2–4 pairs of linear to sword-shaped leaflets plus a terminal one. Pinkish-purple flowers are sparsely arranged on a stalk up to 15cm long. The yellowish-brown, hairy, flattish fruit pods are oblong and slightly sickle-shaped.

Found on sandy flats in the Richtersveld, extending eastwards through Bushmanland to Botswana and northwards into Namibia.
Flowers March to October.

MARYNA KOHRS

RHYNCHOSIA
There are about 65 species of *Rhynchosia* in southern Africa of which three occur in Namaqualand.

Rhynchosia schlechteri
FABACEAE   endemic

A scrambling or spreading shrub up to 1m high and 2m in diameter, the branches and leaves covered with long, outspread, broad-based, glandular, white hairs. The leaves are divided into 3 egg-shaped leaflets; the undersurface is dark-veined, with scattered, yellowish resin dots. The flowers are yellow with brown veins on the large upper petal; the tip of the keel is brown. Fruit pods are obliquely oblong, slightly sickle-shaped and flattish.

Found on rocky slopes in the Richtersveld and Namakwaland Klipkoppe.
Flowers May to October.

OTHOLOBIUM
There are about 45 species of *Otholobium* in southern
Africa of which six occur in Namaqualand.

Otholobium arborescens

vlieëkeurtjie, hook-leaved pea

FABACEAE

A lax, slender shrub up to 4m tall, branching
from near the base; the branches are tan
and rough with soft, grey hairs and distinct
hairless ribbing. The leaves are divided into
3 elliptic leaflets with a sharp, bent tip and
folded slightly inwards; the upper surface
is covered with glands. White to cream-
coloured flowers with a conspicuous purple-
tipped keel are tightly packed at the ends of
branches. Fruit pods are elliptic.

Found on rocky mountain slopes
in the Namakwaland Klipkoppe and
other mountains in the Northern and
Western Cape.
Flowers August to December.

ANNELISE LE ROUX

PSORALEA
There are about 50 species of *Psoralea* in southern
Africa of which one occurs in Namaqualand.

Psoralea glaucescens

FABACEAE

A lax, weeping to erect, much-branched
shrub up to 3m tall. The leaves are divided
into 3 linear leaflets that are shed very
quickly. The greenish-yellow flowers have
a small purple patch on the keel. The small
fruit pods are ovoid.

Found in moist places or along
seasonal streams on mountain slopes in
Namaqualand and on the mountains in
the Northern and Western Cape.
Flowers November to May.

IVAN VAN NIEKERK

ZELDA WAHL

LESSERTIA
There are about 55 species of *Lessertia* in southern Africa of which 12 occur in Namaqualand.

Lessertia diffusa
klappertjie, kloekie, klein-Jantjie-Bêrend

FABACEAE   endemic

A prostrate annual or perennial herb with branches up to 50cm long. The leaves are divided into 11–17 narrow, oblong leaflets, up to 1cm long and sparsely covered with hairs. Pinkish-purple flowers are borne at the end of an elongated, suberect stalk. The egg-shaped to oblong fruit pods are up to 2cm long, inflated, thin and papery; the surface is wrinkled and often marked with purple veins. A palatable plant to stock.

Found in sandy soil on flats on the Coastal Plain and in the Namakwaland Klipkoppe.
Flowers July to September.

HEATHER BURGER

Lessertia spinescens
FABACEAE

A shrub up to 70cm high, with spines 3–9cm long formed by old flower stalks; the branches are minutely and inconspicuously hairy and the older branches are grey. The leaves are divided into 15–21 narrow, oblong leaflets up to 1cm long and sparsely covered with hairs. White and purplish-pink flowers up to 7mm long are borne on a pointed stalk. The almost round fruits are inflated and up to 1.5cm in diameter. A palatable plant to stock.

Found on rocky slopes in the Richtersveld, the Namakwaland Klipkoppe and also in the Cederberg.
Flowers July to October.

Lessertia frutescens
subsp. *frutescens*
(formerly *Sutherlandia frutescens*)
Jantjie-Bêrend, kalkoentjiebos, kankerbos
FABACEAE

A spreading shrub up to 40cm high, with branches that are hairy to almost hairless. The leaves are divided into 13–17 narrowly oblong leaflets up to 1.5cm long and normally hairy below. Red flowers up to 3.5cm long are borne in clusters of 5. Mature fruit pods are inflated, crinkled and less than 2.5 times as long as broad. A palatable plant to stock. This bush is used as a medicine for flu and stomach ailments and is even reputed to cure cancer.

Found on sandy flats and rocky slopes in Namaqualand and in other dry areas in South Africa and Namibia.
Flowers July to November.

ZELDA WAHL

Lessertia frutescens
subsp. *microphylla*
(formerly *Sutherlandia microphylla*)
Jantjie-Bêrend, kalkoentjiebos, kankerbos
FABACEAE

An erect shrub up to 1.5m high, with branches that are often shallowly fluted and usually covered with minute, stiff hairs. The leaves are divided into 17–21 narrowly oblong leaflets up to 1cm long and hairy below. Red flowers up to 3.7cm long are borne in clusters of 2–8. Mature fruits are inflated, smooth and 2.5–4 times as long as broad. A palatable plant to stock. The leaves of this shrub are used as a medicine for flu and stomach troubles and some even believe that it is a cure for cancer.

Found on sandy flats and rocky slopes, often along roadsides, in Namaqualand and in other dry areas in South Africa and Namibia.
Flowers July to December.

ANNELISE LE ROUX

HEATHER BURGER

MURALTIA
There are 118 species of *Muraltia* in southern Africa of which four occur in Namaqualand.

Muraltia spinosa
(formerly *Nylandtia spinosa*)
tortoise berry, skilpadbessie
POLYGALACEAE

A much-branched, erect to spreading shrub up to 1m high, with spine-tipped branches. The blue-green, slightly fleshy leaves are oblong and stalkless. Pale to dark mauve or white flowers, about 5mm long and with a lobed crest in front, are borne singly in leaf axils. The fleshy fruits are round and bright red at maturity. The fruits are edible to humans and palatable to stock.

Found on flats and slopes in Namaqualand, extending southwards into the Western and Eastern Cape. **Flowers June to September.**

ZELDA WAHL

POLYGALA
There are about 80 species of *Polygala* in southern Africa of which five occur in Namaqualand.

Polygala leptophylla
besembos
POLYGALACEAE

An erect, sparse shrub up to 1m high. The hairy leaves are linear-elliptic and sharply tipped. The purple or purplish-pink flowers have a lower petal with a crest that is deeply divided. Fruits are broadly egg-shaped, with a membranous edge.

Found on rocky slopes in the Richtersveld and Namakwaland Klipkoppe, the Cederberg and in southern Namibia. **Flowers August to September.**

CLIFFORTIA
There are about 115 species of *Cliffortia* in southern Africa of which four occur in Namaqualand.

Cliffortia ruscifolia
climbers friend, steekbos

ROSACEAE

An erect shrub up to 1.5m high, with dense, short hairs on the branches. The leathery leaves, borne in clusters, are broad at the base and sharply tipped; the margin is smooth to finely toothed. Flowers are borne singly in leaf axils; male flowers have 12 stamens and the anthers are deep red; female flowers have a deep red stigma.

Found on rocky mountain slopes in the Richtersveld and Namakwaland Klipkoppe and in mountains southwards to Humansdorp.
Flowers September to October.

HEATHER BURGER

FORSSKAOLEA
There are three species of *Forsskaolea* in southern Africa of which two occur in Namaqualand.

Forsskaolea candida
pleisterbos

URTICACEAE

A herbaceous shrub up to 50cm high. Young branches are soft and red, the lower branches woody. Egg-shaped leaves, rounded to pointed, are dark green above and white-woolly below, the margin irregularly toothed. Flowers surrounded by 3–6 green, papery bracts; stalked male flowers form an outer ring around the 1–5 female flowers in the centre. Fruits are egg-shaped, compressed, woolly and enclosed in the bracts.

Found on rocky slopes and in dry riverbeds in the Richtersveld and Nama-kwaland Klipkoppe, extending southwards into the central Karoo and Little Karoo and northwards into Namibia.
Flowers July to November.

ANNELISE LE ROUX

RIAAN DE VILLIERS

PHYLICA
There are about 140 species of *Phylica* in southern
Africa of which seven occur in Namaqualand.

Phylica oleifolia
blinkhardebos
RHAMNACEAE

An evergreen shrub up to 2m tall.
Rigid, wiry branches with very short,
grey hairs. Stalked leaves are elliptic and
pointed; upper leaf surface dark green
and lower surface covered with white
hairs; margin curved slightly backwards.
Male and female flowers borne on
separate plants; cream-coloured flowers,
about 2.5mm long, are covered with white
hairs on the outer surface. Mature, shiny,
reddish, almost round fruits are divided by
3 slight indentations
    Found on rocky slopes in the
Namakwaland Klipkoppe, and from the
Bokkeveld Mountains to Worcester.
Flowers April to June.

NICK HELME

Phylica retrorsa
RHAMNACEAE   endemic

A shrub up to 50cm high; the slender
branches are covered with dense, short,
grey hairs. The small, slightly fleshy, egg-
shaped leaves, borne close together, are
bent backwards. Small flowers covered
with long, silky hairs are borne very close
together at the ends of branches.
    Found on rocky slopes in the
Kamiesberg.
Flowers August to October.

There are about 30 species of *Ficus* in southern Africa of which two occur in Namaqualand.

Ficus cordata
Namaqua fig, Namakwavy
MORACEAE

An upright tree up to 10m tall; the bark is pale grey and smooth; roots grow flattened against rock faces. The hairless leaves are egg-shaped, rounded to heart-shaped at the base and with long, slender stalks. The fruits (figs), brown to purple when mature, are borne in pairs in leaf axils on the branches.

Found among boulders on slopes in Namaqualand, extending further into the Northern and Western Cape and northwards to southern Angola.
Flowers June to January.

ANNELISE LE ROUX

Ficus ilicina
klipvy
MORACEAE

A shrub up to 5m in diameter, spreading flat over the surface of boulders; the branches are furrowed, with yellowish, peeling bark. The elliptic leaves have a prominent midrib below. Round fruits (figs), up to 1cm in diameter and red when mature, are borne in leaf axils.

Growing flat against boulders on slopes in Namaqualand, extending northwards to southern Angola.
Flowers July to December.

TESSA OLIVER

CUCUMIS
There are about 20 species of *Cucumis* in southern Africa of which four occur in Namaqualand.

Cucumis africanus
wild cucumber, wildekomkommer
CUCURBITACEAE

A prostrate perennial herb with hairy branches up to 2m long from a woody tuber; tendrils are found at the nodes. Broadly egg-shaped leaves, up to 10cm long, are deeply 5-lobed and covered with long, stiff hairs. Yellow male flowers borne 5–10 per leaf axil; yellow female flowers borne singly in a leaf axil. Oblong fruits, up to 8cm long, are striped pale greenish-white and purplish-brown; spines stout, laterally compressed and up to 6mm long.

Found in sandy soil on flats, slopes and along watercourses in the Richtersveld and Namakwaland Klipkoppe, and widespread in southern and tropical Africa.
Flowers August to May.

Cucumis rigidus
CUCURBITACEAE

An erect to suberect shrub up to 70cm high; the grey-green to whitish branches do not have tendrils at the nodes. The egg-shaped leaves, up to 5.5cm long, are deeply 3- or 5-lobed. Yellow male flowers up to 7.5mm long are borne 1–4 per leaf axil; yellow female flowers up to 5.5mm long are borne singly in a leaf axil. The elliptic fruits, up to 4.5cm long, are uniform in colour; the spines on the fruit are soft, flattened, up to 6mm long and end in a bulbous-based, short, stiff point.

Found in sandy and rocky flats in the Richtersveld, extending eastwards into Bushmanland and northwards into southern Namibia.
Flowers August to December.

Kedrostis psammophylla
CUCURBITACEAE

A prostrate perennial herb with branches
up to 1m long from a fleshy tuber and
branching underground. The leaves, up
to 25mm long, are 3–5-lobed; the leaf
surface is covered with glandular hairs.
Flowers are borne on the underground
branches and push out of the soil;
greenish male flowers up to 7mm long
are borne in a cluster of up to 25 per
leaf axil; yellow female flowers up to
12mm long are borne singly in a leaf axil.
The round, white fruits develop below
ground, are up to 2.5cm in diameter and
turn green when exposed.

Found on the Coastal Plain, extending
southwards to Redelinghuis.
Flowers May to December.

PIETER VAN WYK

Citrullus lanatus
bitterboela
CUCURBITACEAE

A prostrate annual herb with branches
up to 2m long and robust tendrils at the
nodes. The egg-shaped leaves are up to
20cm long and deeply 3-lobed; the lobes
are further shallowly to usually deeply
incised. Greenish-yellow male and female
flowers up to 3cm in diameter are borne
singly in leaf axils. The mottled fruits are
almost round and up to 20cm in diameter.

Found on sandy flats in the
Richtersveld and Namakwaland Klipkoppe
and widespread in the rest of southern
Africa, tropical Africa and Asia.
Flowers September to December.

PRISCILLA SARGENT

BRAAM VAN WYK

MAYTENUS
There are 13 species of *Maytenus* in southern Africa of which one occurs in Namaqualand.

Maytenus oleoides
klipkershout
CELASTRACEAE

A tree or large shrub up to 3m tall. The leathery, elliptic leaves contain a characteristic pale green sap; the margin is rolled backwards. Small, white flowers flecked with brown are borne in clusters in leaf axils. Fruits are fleshy at first, becoming orange and dry when mature, splitting and thus releasing the seeds.

Found on rocky slopes in the Richtersveld and Namakwaland Klipkoppe, extending eastwards into the mountains of the Northern Cape and southwards to the Cederberg and Grootwinterhoek Mountains.
Flowers April to September.

ANNELISE LE ROUX

GYMNOSPORIA
There are about 30 species of *Gymnosporia* in southern Africa of which three occur in Namaqualand.

Gymnosporia buxifolia
stinkpendoring
CELASTRACEAE

A spreading shrub up to 5m tall; the young branches are brown, green or reddish-purple, becoming grey with age; the branches are spine-tipped. The leaves, borne in clusters, are elliptic with a blunt tip; the margin can be finely and irregularly toothed or entire. Whitish flowers with an unpleasant odour are borne in clusters on short side branches among the leaves. Fruits are thinly leathery to somewhat fleshy, smooth, more or less round and greenish-yellow to yellow, tinged red.

Found on sandy flats and rocky slopes in Namaqualand and widespread in southern and tropical Africa.
Flowers July to April.

OXALIS
There are about 200 species of *Oxalis* in southern
Africa of which 61 occur in Namaqualand.

Oxalis adenodes
suring
OXALIDACEAE

A stemless, deciduous perennial herb up
to 8mm high, with a bulb. The leaves are
divided into 3 leaflets, with the central
leaflet more rounded than the 2 lateral
ones; the surface of the leaflets can be
hairless or glandular-hairy; the leaflet
margin is sometimes fringed with straw-
coloured hairs. White flowers with a
narrow, funnel-shaped, pale yellow tube
are borne singly on a short stalk; the
petals are often spotted with purple on
the outer margin.

Found on rocky slopes and along
streambanks in the Namakwaland
Klipkoppe and in the Hantam Mountains.
Flowers May to August.

ANNELISE LE ROUX

Oxalis ambigua
suring
OXALIDACEAE   endemic

A deciduous perennial herb up to 12cm
high, with a bulb. The numerous hairy
leaves are clustered at ground level or
on a stem up to 5cm high; the leaves are
divided into 3 leaflets, the middle one
tapering to the base and larger than the
2 lateral ones; the leaflets are occasionally
purple-mottled and sprinkled with
numerous blackish dots below; the leaflet
margin is conspicuously hairy. The white,
cream-coloured or yellow flowers have a
widely funnel-shaped, yellow tube; the
petals are purplish near the margin.

Found on rocky slopes in the
Namakwaland Klipkoppe.
Flowers June to July.

ANNELISE LE ROUX

OXALIDACEAE **217**

Oxalis argillacea
suring
OXALIDACEAE

A deciduous perennial herb up to 10cm high, with a bulb. The numerous leaves, borne on a stem up to 3cm long, are clustered together; the leaves are divided into 3 leaflets that have very long, straight, white, glistening hairs on the underside, whereas the insides of the folded leaflets are hairless. The yellow flowers, with calyxes that are densely covered with very long, straight, white, glistening hairs, have a widely funnel-shaped tube; the petals are sometimes purple-margined below and notched.

Found in loamy, gravelly soil on the Knersvlakte, extending southwards to Clanwilliam.
Flowers May to July.

KENNETH OBERLANDER & LÉANNE DREYER

Oxalis blastorrhiza
suring
OXALIDACEAE   endemic

A deciduous, slender perennial herb up to 7cm high, with a bulb. The 6–12 leaves, borne in a cluster on a rigid stem up to 4cm long, are divided into 3 oblong to linear leaflets that are slightly sickle-shaped and folded inwards; the surface of the leaflets is hairy on the underside. Pale lilac to white flowers with a narrow, funnel-shaped, yellow tube are borne on a hairy stalk up to 8mm long; the petals, with violet veins, are long and narrow at the base.

Found in damp soil on the Knersvlakte.
Flowers in July.

KENNETH OBERLANDER & LÉANNE DREYER

Oxalis comosa
bobbejaansuring

OXALIDACEAE   endemic

A deciduous, branched perennial herb up
to 40cm high, with a bulb. The numerous
leaves, borne in clusters at the ends of
branches, are divided into 3 wedge-
shaped leaflets that are themselves
deeply 2-lobed to about halfway. Pale
rose flowers with a short, broadly funnel-
shaped, yellow tube are borne singly on a
stalk up to 10cm long.

Found on rocky slopes, often in shade,
in the Namakwaland Klipkoppe.
Flowers July to September.

ZELDA WAHL

Oxalis dregei
watersuring

OXALIDACEAE

A stemless, deciduous, aquatic perennial
herb up to 10cm high, with a bulb. The
6–12 leaves, loosely overlapping at the
base and borne on stalks up to 5cm long,
are deeply 2-lobed and often folded
inwards. White flowers with a broad,
shallow, yellow cup are borne singly on a
stalk up to 8cm long.

Found in water in seasonal vleis and
along streambanks in the Namakwaland
Klipkoppe and in the Western Cape.
Flowers May to September.

ANNELISE LE ROUX

Oxalis flaviuscula
suring
OXALIDACEAE   endemic

A stemless, deciduous perennial herb
up to 3.5cm high, with a bulb. The 5–27
leaves, borne on a flattened stalk up to
2cm long, are divided into 5–9 linear,
sickle-shaped leaflets that are folded
inwards; the undersurface of the leaflets
is densely covered with hairs. Yellow
flowers with a short, broadly funnel-
shaped tube are borne singly on a stalk
up to 1.5cm long.

Found on sandy flats in the Namakwaland Klipkoppe.
Flowers May to July.

ANNELISE LE ROUX

Oxalis foveolata
suring
OXALIDACEAE   endemic

A stemless, deciduous perennial herb up
to 3cm high, with a bulb. The 6–10 leaves,
borne on very hairy stalks up to 2cm long,
are divided into 3 slightly fleshy leaflets,
the middle one rounder and larger than
the 2 lateral ones; the upper surface of the
leaflets can be hairy or hairless, while the
undersurface is hairy. Yellow flowers with a
cylindrical rather than funnel-shaped tube
are borne singly on a stalk up to 2cm long.

Found on sandy flats in the Richtersveld and Namakwaland Klipkoppe.
Flowers April to May.

ANNELISE LE ROUX

Oxalis furcillata
suring
OXALIDACEAE   endemic

A deciduous perennial herb up to 8cm
high, with a bulb; the plant is stemless or
rarely has a short, sometimes branching
stem up to 1cm long. The numerous basal
leaves, borne on stalks up to 2cm long, are
divided into 3 linear-oblong leaflets that
are folded inwards and deeply indented
into 2 lobes at the tip; the leaflets are
hairless above and densely or sparsely
soft-hairy below. White flowers with a
short, broadly funnel-shaped, yellow tube
are borne singly on a stalk up to 3.5cm
long; the petals are often purple-streaked
towards the outer margin below.
    Found in sandy soil in the Namakwa-
land Klipkoppe.
Flowers May to July.

ANNELISE LE ROUX

Oxalis inconspicua
suring
OXALIDACEAE   endemic

A stemless, deciduous perennial herb up
to 6cm high, with a bulb. The 6–16 basal
leaves, borne on stalks up to 2cm long,
are divided into 3 leaflets that are slightly
indented at the tip; the undersurface of the
leaflets is dotted with many oblong, brown
spots. White, rarely pale pink flowers
are borne on a stalk up to 5cm long; the
slightly inflated, yellow tube is constricted
near the base, with 5 elongated orange
markings at the neck of the tube; the
petals are purple-margined below.
    Found in damp, shady places on
rocky slopes in the Richtersveld and
Namakwaland Klipkoppe.
Flowers May to July.

ANNELISE LE ROUX

Oxalis kamiesbergensis
Kamiesberg-suring

OXALIDACEAE   endemic

A stemless, deciduous perennial herb up
to 3cm high, with a bulb. The 10–30 basal
leaves, borne on stalks up to 1cm long, are
divided into 3 linear-oblong, sickle-shaped
leaflets that are folded inwards; the upper
surface of the leaflets is hairless, while the
lower surface is covered with long, white
hairs. Rose-coloured flowers with a funnel-
shaped, yellow tube are borne on a stalk
about 5mm long.

Found on sandy flats in the Kamiesberg.
Flowers May to June.

Oxalis lichenoides
suring

OXALIDACEAE   endemic

A stemless, deciduous perennial herb
up to 3.5cm high, with a bulb. The basal
leaves, borne on stalks up to 2cm long, are
divided into 3 small leaflets; the leaf stalks
are almost transparent and sometimes
exude a red sap under pressure. White
flowers with a short, broadly funnel-
shaped, yellow tube are borne singly on a
stalk up to 3cm long.

Found in sandy soil on the Knersvlakte.
Flowers June to August

Oxalis louisae
suring
OXALIDACEAE   endemic

A stemless, deciduous perennial herb up
to 14cm high, with a bulb. The 10–30
basal leaves, borne on flattened stalks up
to 4cm long, are divided into 3 leaflets
that are bright green and sometimes
inconspicuously black-dotted. Pale cream-
coloured flowers with a broadly funnel-
shaped, yellow tube are borne singly on a
stalk up to 3.5cm long.
   Found in sandy soil in the Namakwaland Klipkoppe.
Flowers May to June.

Oxalis namaquana
suring
OXALIDACEAE   endemic

A stemless, deciduous perennial herb
up to 15cm high, with a bulb. The basal
leaves, borne on stalks up to 10cm long,
are divided into 3 oblong, bright green
leaflets. Bright yellow flowers with a
short, broad tube are borne singly on a
stalk up to 10cm long.
   Found in seasonally damp places in the
Namakwaland Klipkoppe.
Flowers May to September.

ANNELISE LE ROUX

Oxalis obtusa
suring
OXALIDACEAE

A deciduous perennial herb up to 15cm high, with a bulb; the plant can be stemless or have a short, hairy stem if growing in shady areas; all parts of the plant are covered with hairs that bend downwards. The 10–20 leaves, crowded together at the same point on the stem and borne on stalks up to 4cm long, are divided into 3 oblong leaflets that are indented at the tip. White, pale pink to orange-pink, brick-red or yellow flowers with a short, funnel-shaped, yellow tube are borne singly on a stalk up to 10cm long.

Found in sandy or loamy soil in Namaqualand, Bushmanland, the Roggeveld Mountains, the Western and Eastern Cape and Namibia. **Flowers June to October.**

ZELDA WAHL

Oxalis pes-caprae
langbeensuring
OXALIDACEAE

A deciduous perennial herb up to 40cm high, with a bulb; the leaves are borne in a rosette perched on a stem up to 20cm long. The 10–40 leaves, borne on stalks up to 12cm long, are divided into 3 broad, wedge-shaped leaflets that are slightly 2-lobed. Yellow flowers with a broad, funnel-shaped tube are borne in a cluster at the top of a stalk up to 30cm long. The leaves and flowers are cooked with goat's milk to make a tasty porridge. A palatable plant to stock in winter.

Found in sandy soil, often among rocks in Namaqualand, extending into the Western and Eastern Cape and into Namibia. **Flowers May to August.**

Oxalis purpurea
suring
OXALIDACEAE

A stemless, deciduous perennial herb up to 15cm high, with a bulb. The leaves, lying flat on the ground and borne on round or flat, slightly fleshy stalks, are divided into 3 leaflets that are often deep purple below. The flowers, rose-purple, deep rose, various shades of pale violet, salmon, yellow, cream-coloured or white, have a cup-shaped, yellow tube.

Found on flats and slopes in the Namakwaland Klipkoppe and on the Knersvlakte, extending into the Western and Eastern Cape.
Flowers May to September.

ANNELISE LE ROUX

Oxalis sonderiana
suring
OXALIDACEAE

A stemless, deciduous perennial herb up to 4cm high, with a bulb. The 6–10 blue-green, leathery, basal leaves, borne on stalks up to 2cm long, are divided into 3 leaflets; the leaflet margin is minutely glandular-hairy. Yellow or white flowers with a funnel-shaped, yellow tube are borne singly on a stalk up to 2cm long.

Found on rocky slopes in Namaqualand, extending southwards to Graafwater.
Flowers April to June.

ANNELISE LE ROUX

ANNELISE LE ROUX

RICINUS
There is only one exotic species of *Ricinus* in southern Africa and in Namaqualand.

Ricinus communis
castor oil bean, kasterolieboom

EUPHORBIACEAE   alien invader

A shrub up to 3m tall. The broadly egg-shaped leaves, up to 30cm in diameter, are deeply lobed into 5–9 segments; they give off a pungent odour when touched. Female flowers are borne at the top of branches; male flowers are borne below them. The round, green to brown fruits, up to 2cm in diameter, are covered with soft spine-like projections. An invader, originally from northeastern Africa and India. Medicinal castor oil is made from the seed after removal of the toxic components.

Found in watercourses or in disturbed areas in Namaqualand and throughout southern Africa.
Flowers August to January.

ANNELISE LE ROUX

EUPHORBIA
There are about 210 species of *Euphorbia* in southern Africa of which 39 occur in Namaqualand.

Euphorbia dregeana
diklootmelkbos, dikboudmelkbos

EUPHORBIACEAE

An erect, deciduous shrub up to 1.5m, branching from the base and secreting a milky sap when damaged. The slightly fleshy, pale greyish-green, cylindrical branches, up to 2.5cm in diameter, have prominent leaf scars. Leaves are egg-shaped, pointed, up to 8mm long and 5mm wide. Yellow flowerheads are borne in loose groups at the ends of branches. The round, minutely hairy fruits are up to 1cm in diameter.

Found on gravelly flats and rocky slopes in the Richtersveld and Namakwaland Klipkoppe, extending eastwards into Bushmanland and northwards into southern Namibia.
Flowers April to September.

Euphorbia ephedroides
steenbokmelkbos
EUPHORBIACEAE

An erect, deciduous shrub up to 50cm high, branching from the base and secreting a milky sap when damaged. The fleshy, yellow-green to grey-green, cylindrical branches are up to 4mm in diameter. The leaves, up to 1cm long and 4mm across although often much smaller, are oblong to elliptic. Yellow flowerheads are borne in clusters of 3 at the ends of branches. The round fruits are up to 4mm in diameter.

Found on sandy flats and rocky slopes in Namaqualand, extending eastwards to Pofadder and northwards into southern Namibia. **Flowers July and October.**

ANNELISE LE ROUX

Euphorbia gariepina
EUPHORBIACEAE

A spreading, much-branched, deciduous shrub up to 75cm high, secreting a milky sap when damaged. The fleshy, pale grey, cylindrical branches taper towards the tips and are covered with scattered, conical tubercles; the ends of the branches dry out but do not become spiny. The bluish-green, elliptic, sharply pointed leaves, borne on a stalk up to 1cm long, are up to 1.5cm long, up to 5mm across and are folded inwards. Yellow flowerheads are borne in clusters of 3 at the ends of branches. The round fruits are up to 6mm in diameter.

Found on rocky flats and lower rocky slopes in the Richtersveld, extending eastwards to Prieska and northwards to Angola. **Flowers August and October.**

NORBERT JÜRGENS

NORBERT JÜRGENS

Euphorbia gregaria
EUPHORBIACEAE

An erect, deciduous shrub up to 1m high, branching from the base and secreting a milky sap when damaged. The fleshy, cylindrical branches are blue-green. Small, yellow flowerheads are borne in clusters at the ends of branches. The drooping fruits, borne on reddish stalks up to 1.5cm long, are up to 2cm in diameter and covered in minute, tawny hairs.

Found on sandy to gravelly flats in the Richtersveld, extending eastwards to Kakamas and northwards into southern Namibia.
Flowers July to November.

ANNELISE LE ROUX

Euphorbia hamata
beesmelkbos, beeskrag, olifantsmelkbos
EUPHORBIACEAE

A spreading, deciduous shrub up to 45cm high, branching at the base and secreting a milky sap when damaged. The fleshy, green to reddish, obscurely 3-angled branches, up to 1.5cm in diameter, have prominent conical-horizontal or recurved tubercles. The slightly fleshy, stalkless, egg-shaped, sharply pointed leaves are up to 2cm long, 1cm across and are folded slightly inwards. Flowerheads are borne singly at the ends of branches. The round fruits are up to 6mm in diameter. Cattle seem to show a preference for this species.

Found on flats and rocky slopes in Namaqualand, extending from Namibia southwards to Worcester.
Flowers April to October.

228 EUPHORBIACEAE

Euphorbia mauritanica
gifmelkbos
EUPHORBIACEAE

An erect, much-branched, deciduous shrub
up to 1m high, secreting a milky sap when
damaged. The fleshy, cylindrical, yellow-
green branches, up to 1cm in diameter,
leave scattered scars on the older
branches. The leaves, up to 1.3cm long
and 5mm across, are sword-shaped. Yellow
flowerheads are borne in clusters at the
ends of branches. The round fruits are up
to 6mm in diameter. This plant is reputed
to be poisonous and only steenbok and
klipspringer are known to eat it.

Found on sandy and rocky flats,
sometimes on lower rocky slopes in
Namaqualand, extending from Angola to
the Cape Peninsula to KwaZulu-Natal.
Flowers July to September.

ZELDA WAHL

Euphorbia muricata
EUPHORBIACEAE

A spreading, much-branched, deciduous
shrub up to 45cm high, secreting a milky
sap when damaged. The fleshy, grey-
green branches, up to 1cm in diameter,
are covered with laterally compressed
ridges and warts; the ends of the branches
sometimes die off but do not become
spiny. The triangular leaves are up to 2mm
long and recurved. Yellow flowerheads
are borne in clusters at the ends of
branches. The round fruits are up to 3mm
in diameter.

Found on sandy flats and gentle, rocky
slopes, often among white quartz pebbles,
on the Coastal Plain and Knersvlakte,
extending eastwards to Calvinia.
Flowers July to September.

ADRIAN FORTUIN

NORBERT JÜRGENS

Euphorbia phylloclada
bontmelkbos
EUPHORBIACEAE

A spreading or prostrate annual herb up to 15cm high, secreting a milky sap when damaged. The egg-shaped, sharply pointed, blue-green leaves, up to 2mm long and 2cm across, are crowded at the ends of short branches; the leaves are green at the tips and white at the base; the margin is white, yellow or reddish. Yellow flowerheads are hidden among the leaves. The drooping fruits are borne on a stalk, protruding above the leaves.

Found on rocky slopes and in watercourses in the Richtersveld, extending northwards into southern Namibia. **Flowers July to September.**

ZELDA WAHL

Euphorbia rhombifolia
(formerly *Euphorbia decussata*)
soetmelkbos
EUPHORBIACEAE

An erect to spreading, rigid, deciduous shrub up to 60cm high, secreting a milky sap when damaged. The slightly fleshy, blue-green, cylindrical branches are up to 8mm in diameter; the stems branch dichotomously at an angle of 90–120°; the ends of the branches rapidly die back to form spines. The small, rudimentary triangular leaves drop off early. Yellow flowerheads are borne in sparse clusters at the ends of branches. The round fruits, borne erect or hanging, are up to 4.5mm in diameter. A palatable to very palatable plant to stock.

Found on rocky flats and slopes in Namaqualand, extending from Namibia to the Free State and Eastern Cape. **Flowers March to October.**

Euphorbia stapelioides
EUPHORBIACEAE

A spreading to prostrate, deciduous dwarf
succulent up to 10cm high, with a tuber
and branching rhizomes, and secreting
a milky sap when damaged. The fleshy,
green to reddish, cylindrical branches are
up to 10cm long and 1.5cm in diameter.
The fleshy, stalkless leaves are oblong to
triangular with recurved tips. Flowerheads
are borne at the ends of branches. The
round fruits are hairy to hairless and
about 3mm in diameter.

Found on sandy or stony flats, often
among quartz pebbles, on the Coastal
Plain, extending northwards into
southern Namibia.
Flowers August to September.

CARLO VAN TONDER

Euphorbia virosa
boesmangif
EUPHORBIACEAE

A deciduous, spiny succulent up to 2.5m
tall, secreting a milky sap when damaged.
The 5–8-angled stems, up to 9cm in
diameter, are constricted into segments
at irregular intervals; pairs of spines, 1cm
long, are united by a continuous horny
margin along the angles. The triangular
to egg-shaped leaves, up to 3mm long,
are shed very quickly. Yellow flowerheads
are borne along the angles at the ends
of branches. Fruits are round to 3- or
4-angled and up to 2.5cm in diameter.

Found on dry rocky slopes close to
the Orange River in the Richtersveld,
extending northwards to Angola.
Flowers August to October.

ANNELISE LE ROUX

MARYNA KOHRS

Boscia albitrunca
shepherd's tree, witgatboom
CAPPARACEAE

An evergreen tree or large shrub up to 2m tall. The trunks have a smooth, whitish bark; the branches sometimes droop when young but become stiff with age, but are usually not spiny. The elliptic, leathery, grey-green leaves are up to 5cm long and 1cm across. Flowers are clustered together and have no petals, but many yellow-green stamens protrude. The hairless, round fruits are up to 1.5cm in diameter.

Found on rocky slopes in the Richtersveld, Namakwaland Klipkoppe and also widespread in dry areas in southern Africa.
Flowers August to October.

ANNELISE LE ROUX

Cadaba aphylla
desert spray, swartstorm
CAPPARACEAE

A tangled shrub up to 1.5m high. Blue-green branches are usually sharply pointed. Leaves only present on seedlings and some very young branches. Flowers without petals are borne in a flat cluster; yellowish, orange to reddish-purple sepals covered with many sticky glands; red stamens fused in a long tubular column; ovary borne on the style projecting beyond the staminal column. The cylindrical fruits are covered with sticky glands. A palatable species.

Found on sandy flats, rocky slopes and in dry riverbeds in Namaqualand and also widespread in dry areas in southern Africa.
Flowers August to March.

Cleome foliosa var. *lutea*
CLEOMACEAE

An annual or perennial herb up to 50cm high. Branches and leaves may be hairless or covered with glandular hairs. The elliptic leaves are divided into 3 leaflets that are folded slightly inwards. Yellow flowers are borne at the end of a branch; the 20–30 stamens are unequal in length, with 5 or 6 longer and stouter than the rest, exceeding the petals. The thin, cylindrical fruits are up to 8cm long.

Found on sandy flats and in dry watercourses in the Richtersveld, extending eastwards into Bushmanland and northwards into southern Namibia. **Flowers March to September.**

ANNELISE LE ROUX

Heliophila amplexicaulis
white sporrie, witsporrie
BRASSICACEAE

An erect annual herb up to 40cm high. The leaves, up to 4cm long and 1.5cm across, are sword-shaped, with the base clasping the stem. White, mauve or pink flowers up to 1cm in diameter are borne at the ends of branches. The drooping fruits, up to 3.5cm long and 3mm in diameter at the broadest parts, are angled downwards and look like a string of beads.

Found on sandy flats in the Namakwa-land Klipkoppe and the Calvinia area southwards to Laingsburg. **Flowers August to October.**

ZELDA WAHL

ZELDA WAHL

Heliophila arenaria
blue sporrie, blousporrie
BRASSICACEAE

An erect annual herb up to 50cm high, with hairy branches. The linear leaves, up to 12cm long, are simple or lobed, with or without hairs. Blue flowers with a white centre are borne at the ends of branches. The drooping fruits, up to 5cm long and 2mm in diameter, have straight sides or look like a string of beads.

Found in sandy soil in the Namakwa-land Klipkoppe and on the Knersvlakte, extending southwards to Piketberg.
Flowers July to September.

ANNELISE LE ROUX

Heliophila carnosa
blompeperbossie
BRASSICACEAE

An erect perennial shrub up to 50cm high. The leaves, crowded near the woody base of the plant, are thread-like to strap-shaped or lobed, often slightly fleshy with sharp tips. White, pale mauve, pink, purple or blue flowers are borne at the ends of erect stalks up to 40cm long. The drooping fruits are broadly linear with straight margins.

Found on rocky slopes in the Richtersveld, the Namakwaland Klip-koppe and also throughout the rest of South Africa.
Flowers August to October.

Heliophila coronopifolia
blue sporrie, blousporrie
BRASSICACEAE

An erect annual herb up to 60cm high. The linear leaves, up to 10cm long, are deeply lobed into narrow segments. Pale to bright blue flowers with a lighter centre are clustered at the ends of branches. The drooping fruits, up to 8cm long and 2mm in diameter, look like a string of beads.

Found on flats and slopes in Namaqualand, extending southwards to Caledon.
Flowers August to October.

ZELDA WAHL

Heliophila juncea
(formerly *Brachycarpaea juncea*)
bergriet, ridderspoor
BRASSICACEAE

A lax, tufted shrub up to 70cm high. The leaves, up to 7mm broad, are linear to narrowly oblong and sharply pointed. Mauve, pink, blue or white flowers about 3cm in diameter are borne loosely at the ends of branches. The erect or recurved, woody fruits are almost round.

Found on rocky slopes in the Namakwaland Klipkoppe and also in other mountains in the Western Cape.
Flowers August to December.

HEATHER BURGER

BRASSICACEAE **235**

ZELDA WAHL

Heliophila laciniata
basket sporrie, mandjiesporrie
BRASSICACEAE   endemic

An annual herb up to 50cm high. The basal
leaves, up to 15cm long, are lobed into 5–14
narrow segments; the leaves are crowded
in a basal rosette, giving the appearance
of a basket. White flowers, about 2cm in
diameter and becoming mauve with age,
are borne in loose clusters at the end of a
stalk. The drooping fruits, up to 4cm long
and 3mm broad, have straight margins.
   Found on sandy flats and in dry
watercourses in the Namakwaland
Klipkoppe.
Flowers August to September.

ZELDA WAHL

Heliophila lactea
blue sporrie, blousporrie
BRASSICACEAE

An erect annual herb up to 40cm high.
The linear leaves, up to 9cm long and
1mm across, are unlobed. Pale blue or
mauve flowers with a white centre and
up to 1.5cm in diameter are loosely
clustered at the end of a stalk. The slightly
drooping, cylindrical fruits are up to 6cm
long and 2mm in diameter.
   Found in sandy soil in Namaqualand,
extending eastwards to Calvinia,
southwards to Colesberg and northwards
into southern Namibia.
Flowers July to October.

Heliophila trifurca
beessporrie
BRASSICACEAE

An annual herb up to 60cm high. The
grey-green, linear leaves, up to 12cm long
and 2.5mm across, are unlobed to 5-lobed
with very narrow segments. Mauve or
white flowers up to 1.5cm in diameter are
borne at the ends of branches. Fruits are
borne horizontally, have straight margins
and are up to 2.5cm long and 4mm broad.
When cows eat this plant, the flavour of
their milk becomes tainted.

Found on sandy flats or in dry
watercourses in the Richtersveld,
extending eastwards into Bushmanland,
southwards to Laingsburg and northwards
into southern Namibia.
Flowers March to September.

ZELDA WAHL

Heliophila variabilis
sporrie
BRASSICACEAE

An erect annual herb up to 35cm high.
The leaves, up to 7cm long, are lobed
into 1–3 pairs of widely spaced, linear
segments. Flowers up to 1.3cm in diameter
are borne in loose clusters at the ends
of branches, their colour changing from
white to mauve or lilac with age. Fruits up
to 4cm long and 3mm broad are slightly
beaded and are borne upright.

Found on rocky slopes in Namaqualand,
extending southwards to Laingsburg.
Flowers July to September.

ZELDA WAHL

BRASSICACEAE **237**

There are about 35 species of *Commiphora* in southern Africa of which four occur in Namaqualand.

Commiphora capensis
Namaqua corkwood, Namakwa-kanniedood
BURSERACEAE

A shrub up to 1m high, with a thick trunk branching repeatedly above ground level. The bark, brown to green with blackish patches, peels into small, white, papery pieces. The leaves are divided into 3 leaflets; the leaflet margin is coarsely sinuate. Small, yellow to green male and female flowers are borne on separate plants. The elliptic fruits are flattened.

Found on rocky slopes in the Richtersveld, extending eastwards into Bushmanland and northwards into southern Namibia.
Flowers October to March.

ANNELISE LE ROUX

Commiphora namaensis
Nama corkwood, Nama-kanniedood
BURSERACEAE

A shrub up to 75cm high, with thick branches just above ground level and thinner branches higher up. The pale grey bark does not peel. The leaves are orbicular or slightly oblong; the margin is toothed. Cream-coloured male and female flowers are borne on separate plants. Fruits are almost round.

Found on rocky slopes in the Richtersveld, extending northwards into southern Namibia.
Flowers September to November.

ANNELISE LE ROUX

OZOROA
There are about 20 species of *Ozoroa* in southern
Africa of which three occur in Namaqualand.

Ozoroa concolor
green resin tree, groenharpuisboom
ANACARDIACEAE

A deciduous tree up to 4m tall, with grey
bark. The elliptic, leathery, shiny, dark
green leaves, crowded at the ends of the
branchlets, are up to 5cm long and 2.5cm
across. Creamy-yellow, sweetly scented
flowers are borne in clusters at the ends of
branches. Fruits are broader than long and
become reddish when mature.

Found along dry watercourses in the
Richtersveld, extending northwards into
southern Namibia.
Flowers September to October.

PIETER VAN WYK

Ozoroa dispar
kliphout, t'ôrrieboom
ANACARDIACEAE

An evergreen tree up to 4m tall, with
grey bark. The elliptic, leathery, bright
green leaves, crowded at the ends of
the branchlets, are up to 5cm long and
3cm across; the leaf is strongly and
conspicuously yellow-veined. Creamy-
white flowers are borne in clusters at the
ends of branches. Fruits are kidney-shaped
and become black when mature.

Found on rocky slopes in the
Richtersveld and Namakwaland Klipkoppe,
extending northwards into Namibia.
Flowers May.

ANNELISE LE ROUX

ANACARDIACEAE **239**

There are about 75 species of *Searsia* in southern Africa of which seven occur in Namaqualand.

Searsia horrida
(formerly *Rhus horrida*)
rooidoring
ANACARDIACEAE   endemic

A much-branched, spreading, evergreen shrub up to 1.5m high, with rigid, spiny branches; the new growth is reddish-brown. The dull grey-green leaves (reddish-brown when young), borne on a leafy stalk, are divided into 3 leaflets, the terminal one larger than the lateral ones; the surface of the leaflets is densely covered with minute hairs. Greenish-yellow flowers are crowded on dwarf outgrowths of often leafless spines. The asymmetrically oblong fruits are dark reddish-brown, drying to black.

Found on rocky slopes in the Namakwaland Klipkoppe. **Flowers May to July.**

Searsia incisa
(formerly *Rhus incisa*)
kalwertaaibos
ANACARDIACEAE

A much-branched, deciduous shrub up to 1.5m high. The leaves are divided into 3 egg-shaped to inversely egg-shaped leaflets, the terminal one longer than the lateral ones; the surface of the leaflets is minutely hairy above and covered with many white-woolly hairs below; the leaflet margin is shallowly toothed. Yellow-green flowers are borne in clusters. The light brown fruits are thickly covered with hairs. A palatable plant to stock.

Found in the Richtersveld and Namakwaland Klipkoppe, extending into the Western and Eastern Cape. **Flowers June to August.**

Searsia longispina
(formerly *Rhus longispina*)
doringtaaibos
ANACARDIACEAE

A much-branched, evergreen shrub up
to 3m tall, with pale, spiny branches. The
leaves are divided into 3 inversely egg-
shaped leaflets, indented at the tip, the
terminal one larger than the lateral ones.
Pale yellow, sweetly scented flowers are
borne in clusters at the ends of branches.
The shiny, brown fruits are lens-shaped.
    Found in deep sand on the Coastal
Plain, extending to KwaZulu-Natal.
Flowers May to November.

PAUL KRUGER

Searsia populifolia
(formerly *Rhus populifolia*)
suurtaaibos
ANACARDIACEAE

A much-branched, deciduous shrub up to
2.5m tall. The leaves, borne on a slightly
furrowed and winged stalk, are divided
into 3 widely elliptic leaflets; the surface
of the leaflets is dark green, shiny, hairless
above and covered with yellow hairs
below; the terminal leaflet is larger than
the lateral ones; the leaflet margin is
scalloped. Greenish-yellow flowers are
borne in clusters at the ends of branches.
Fruits are asymmetrical, lens-shaped and
have 3 small spines at the top.
    Found on hills in the Richtersveld
and Namakwaland Klipkoppe, extending
eastwards into Bushmanland and northwards into Namibia.
Flowers May to January.

ZELDA WAHL

ANACARDIACEAE **241**

Searsia undulata
(formerly *Rhus undulata*)
taaibos, t'narra
ANACARDIACEAE

A much-branched, evergreen shrub up to 2m tall. The leaves, borne on a furrowed stalk, are divided into 3 inversely egg-shaped, glossy, sticky leaflets, the terminal one longer than the lateral ones; the leaflet margin is slightly undulating. Small, yellow flowers are borne in clusters at the ends of branches. The round fruits are hairless, shiny and dull yellow to cream-coloured, often reddish on the exposed side. An unpalatable plant to stock except for the fruits.

Found on rocky slopes in Namaqualand, extending southwards to Ladismith and northwards into southern Namibia. **Flowers April to June.**

ZELDA WAHL

ANNELISE LE ROUX

DODONAEA
There is only one species with two varieties of *Dodonaea* in southern Africa of which only one variety occurs in Namaqualand.

Dodonaea viscosa
var. *angustifolia*
fever tree, ysterhout, t'oubee
SAPINDACEAE

A tree or large shrub up to 5m tall. The stalked, long, shiny, pale green, narrowly elliptic leaves are resinous. Greenish-yellow flowers are borne in small clusters between the leaves or at the ends of branches; male and female flowers are borne on separate plants. Fruits are 3-winged and papery when mature.

Found at the bottom of large granite rock sheets in the Richtersveld and Namakwaland Klipkoppe and on level areas in the Kamiesberg, extending into the Western and Eastern Cape. **Flowers April to October.**

ERYTHROPHYSA
There are two species of *Erythrophysa* in southern
Africa of which one occurs in Namaqualand.

Erythrophysa alata
mannetjie-roldoring, kloe-kloekie

SAPINDACEAE

A shrub or tree up to 3m tall. The leaves,
up to 10cm long, are divided into 4 pairs
of opposite leaflets plus a terminal one;
the midrib is conspicuously winged. The
red flowers are about 2cm in diameter;
the stamens are much longer than the
petals. The reddish fruits, inflated and
bladder-like, are divided into 3 sections.
The round seeds, about 1.5cm in diameter,
are black and very hard. A palatable plant
to stock.

Found on rocky slopes in the Namakwaland Klipkoppe and in Namibia.
Flowers March to July.

IVAN VAN NIEKERK

NYMANIA
There is only one species of *Nymania* in southern
Africa and in Namaqualand.

Nymania capensis
klapperbos

MELIACEAE

An erect, evergreen shrub up to 2m tall.
The leathery, narrowly linear-elliptic leaves
are up to 3cm long and 5mm across. Pale
pink to red flowers about 5cm in diameter
are borne in leaf axils. The cream-coloured
to pink fruits are inflated papery capsules
divided into 4 sections. A palatable plant
to stock.

Found on flats, rocky slopes and in
dry riverbeds in the Richtersveld and the
Namakwaland Klipkoppe, extending into
Bushmanland, the Western and Eastern
Cape and Namibia.
Flowers May to November.

ZELDA WAHL

AGATHOSMA
There are about 140 species of *Agathosma* in
southern Africa of which two occur in Namaqualand.

Agathosma capensis
anysboegoe
RUTACEAE

An evergreen shrub up to 75cm high. The
linear leaves, up to 7mm long, can be
hairy or hairless; the leaf surface is covered
with translucent dots; crushed leaves are
sweetly spice-scented. White, pink or
mauve flowers up to 5mm long are borne
in round clusters at the ends of branches.
    Found in sandy soil on the Coastal
Plain, extending into the Western and
Eastern Cape.
Flowers August to October.

RIAAN DE VILLIERS

DIOSMA
There are about 30 species of *Diosma* in southern
Africa of which two occur in Namaqualand.

Diosma acmaeophylla
ribbokboegoe, valsboegoe
RUTACEAE

An evergreen shrub up to 2m tall. The
slightly fleshy, linear leaves are arched,
tapering to a sharp, curved tip; the leaf
surface is covered with translucent dots;
crushed leaves are penetratingly scented.
The flowers are white to cream-coloured.
Fruits are 5-horned.
    Found on rocky slopes in the Nama-
kwaland Klipkoppe and on mountains
extending southwards as far as
Matjiesfontein.
Flowers March to November.

ANNELISE LE ROUX

Diosma ramosissima
RUTACEAE

An evergreen shrub up to 1m high. The slightly fleshy, dull blue-green leaves are straight, seldom longer than 4mm, usually with a blunt tip, and have gland dots in rows on either side of the midrib; the crushed leaves have a penetrating scent. White flowers are borne in loose clusters at the ends of branches. Fruits are 5-horned.

Found on sandy flats and rocky mountain slopes in the Namakwaland Klipkoppe and on the Coastal Plain, extending southwards to Worcester. **Flowers August to November.**

NICK HELME

CYTINUS
There are three species of *Cytinus* in southern Africa of which one occurs in Namaqualand.

Cytinus sanguineus
aardroos
CYTINACEAE

A root parasite without leaves. The funnel-shaped flowers are bright red to dark pink; the staminal column protrudes from the male flowers.

Found in sandy soil, under bushes or in the open, in Namaqualand, extending into the Northern, Western and Eastern Cape. **Flowers May to December.**

♀
♂

ELIZE HOUGH

HERMANNIA
There are about 145 species of *Hermannia* in southern
Africa of which about 27 occur in Namaqualand.

Hermannia amoena
jeukbos
MALVACEAE

An erect shrub up to 1m high. The bluish-
green, egg-shaped to oblong leaves, up to
4cm long and 2cm across, are rounded at
the base and tip and are corrugated when
young, with prominent veins below; the
surface of the leaf is covered with minute,
star-like hairs; the margin is scalloped.
Yellow flowers turning orange-red with
age are borne in loose clusters towards
the ends of branches.

Found on sandy flats or rocky slopes in
Namaqualand, extending northwards into
southern Namibia.
Flowers August to September.

ZELDA WAHL

Hermannia cuneifolia
broodbos
MALVACEAE

A branched shrub up to 50cm high. The
wedge-shaped leaves, up to 2cm long
and 1cm across, are narrowed towards
the base and coarsely toothed above; the
surface of the leaf is covered with minute,
star-like hairs. The sweetly scented flowers
are yellow and become brown with age. A
very palatable plant to stock.

Found on rocky slopes in
Namaqualand, extending into the Western
and Eastern Cape, the southern Free State
and Lesotho.
Flowers August to September.

ZELDA WAHL

Hermannia disermifolia
jeukbos
MALVACEAE

An erect branched shrub up to 1m high.
The long-stalked, yellow-green, oblong
leaves, up to 3cm long and 1cm across, are
deeply corrugated, even when older, with
prominent veins below; the surface of the
leaf is covered with minute, star-like hairs;
the margin is crisped. Yellow flowers are
borne towards the ends of branches.

　　Found on rocky slopes in the
Richtersveld and Namakwaland Klipkoppe,
extending eastwards into Bushmanland.
Flowers August to September.

ANNELISE LE ROUX

Hermannia macra
MALVACEAE

A slender, erect shrub up to 50cm high.
The oblong leaves, borne mostly at the
woody base of the plant, are coarsely and
unevenly lobed, but never to the midrib;
the leaves have 3 prominent veins; the
surface of the leaf is covered with minute,
star-like hairs. Yellow to orange-red
flowers are borne on a much-branched
stalk up to 45cm long.

　　Found on sandy flats and in dry
riverbeds in Namaqualand, extending
eastwards into Bushmanland and northwards into Namibia.
Flowers August to October.

ALTA CHAMBERS

MALVACEAE **247**

Hermannia meyeriana
MALVACEAE   endemic

A prostrate, deciduous perennial herb with branches up to 35cm long from a woody rootstock. The grey-green leaves are doubly and deeply lobed; the surface of the leaf is covered with minute, star-like hairs. The hanging, cup-shaped flowers are pink with red central bands.

Found on sandy flats in the Richtersveld and Namakwaland Klipkoppe.
Flowers September to October.

VERONICA ESTERHUIZEN

Hermannia muricata
MALVACEAE

A shrub up to 50cm high, often much-branched from the base. The narrowly oblong leaves, up to 3.5cm long and 1cm across, are rounded at the base, with prominent, indented veins; the surface of the leaf is covered with minute, star-like hairs; the margin is sinuate and slightly recurved. Hanging, yellow flowers are borne on long, lax branches.

Found on rocky slopes in the Kamiesberg, in the mountains from Calvinia to Citrusdal and in the southern Karoo to George.
Flowers September to November.

HEATHER BURGER

Hermannia stricta
desert rose, woestynroos
MALVACEAE

A perennial shrub with slender, erect
branches up to 80cm high. The leaves are
inversely sword-shaped; the surface of
the leaf is covered with minute, star-like
hairs; the margin is toothed, especially at
the end. The drooping, dark dusty pink
flowers are up to 3cm long; the petals
overlap each other spirally. A palatable
plant to stock.

Found on sandy flats and rocky slopes
in the Richtersveld, extending eastwards
into Bushmanland and northwards into
southern Namibia.
Flowers January to December after rain.

HEATHER BURGER

Hermannia trifurca
broodbos, koerasie
MALVACEAE

A much-branched shrub up to 75cm high.
The oblong leaves, up to 1.5cm long, have
a prominent midrib below; the surface of
the leaf is thinly covered with minute, star-
like hairs; the margin has 3 teeth at the tip.
The bell-shaped flowers, turned to one side
at the ends of branches, range from pale
pink to purple all on the same branch. A
very palatable plant to stock.

Found on sandy flats and rocky slopes
in Namaqualand, extending southwards
into the Western Cape and northwards
into southern Namibia.
Flowers August to September.

ANNELISE LE ROUX

MALVACEAE **249**

RADYERA
There is only one species of *Radyera* in southern Africa and in Namaqualand.

Radyera urens
sandpampoentjie, wildepampoen
MALVACEAE

A prostrate, deciduous perennial herb with branches up to 1m long from a woody rootstock; the whole plant is covered with dense, irritant, star-like hairs. The kidney-shaped leaves, up to 12cm across, are prominently veined below; the margin is scalloped. Orange-red to brownish flowers about 4cm long are borne at ground level. The roots of this plant are cooked and used as a remedy for piles.

Found on sandy flats in Namaqualand and other dry areas of southern Africa. **Flowers October to March.**

ANISODONTEA
There are about 20 species of *Anisodontea* in southern Africa of which three occur in Namaqualand.

Anisodontea bryoniifolia
wildestokroos
MALVACEAE

An erect, slender shrub up to 2m tall, with hairy, yellow young branches. The egg-shaped leaves, up to 4cm long and across, are divided into 3 lobes, the middle one the largest; the leaf surface is covered with hairs; the margin is crisped. The flowers are pale pink, with crimson markings at the base on the inside; the stamens form a central column.

Found on rocky mountain slopes, often near streams, in the Richtersveld and Namakwaland Klipkoppe and in the mountains from the Cederberg to Tulbagh. **Flowers January to December after rain.**

Struthiola ciliata
jakkalsgare

THYMELAEACEAE

An erect, slender shrub up to 1m high, with branches slightly square in cross-section and covered with white-woolly hairs. The linear to sword-shaped to narrowly elliptic leaves are concave above and convex below; the margin and tip have white hairs. Pale yellow flowers, up to 2cm long and with long hairs surrounding the petals, are borne in clusters along the branches. The bark tears off in long strips when one tries to break off a twig.

Found on slopes in the Namakwaland Klipkoppe and in mountains in the Western Cape and the Eastern Cape. **Flowers January to December after rain.**

RIAAN DE VILLERS

Struthiola leptantha
jakkalsgare, kannabas

THYMELAEACEAE

An erect, slender shrub up to 1m high, with hairy branches slightly square in cross-section. The narrowly elliptic leaves, up to 1cm long, are flat above and convex below; the margin is hairy when young, becoming almost smooth later. Pale yellow flowers, up to 2cm long and with short hairs surrounding the petals, are borne at the ends of branches. The bark tears off in long strips when one tries to break off a twig.

Found on sandy flats or rocky slopes in Namaqualand, extending to Malmesbury and in the Little Karoo. **Flowers June to October.**

ANNELISE LE ROUX

THYMELAEACEAE **251**

ZELDA WAHL

GRIELUM
There are five species of *Grielum* in southern Africa of
which four occur in Namaqualand.

Grielum grandiflorum
platdoring
NEURADACEAE

A prostrate, deciduous perennial herb
with creeping branches up to 35cm long
and a long, thick, woody root. The hard
leaves are doubly lobed into narrow,
pointed segments; the leaf surface is
covered with silver hairs; the margin is
rolled under. The sulphur-yellow flowers
are up to 5.5cm in diameter. The flat fruits
have 5 horny spikes and woody lobes.

   Found on sandy flats on the Coastal
Plain, extending to the Cape Peninsula.
Flowers July to October.

TED OLIVERL

Grielum humifusum
pietsnot, duikerwortel, t'koeibee
NEURADACEAE

A prostrate annual herb with creeping
branches up to 35cm long from a fleshy
root. The soft leaves are lobed into broad
segments; the underside of the leaf is
covered with white hairs. The brilliant
lemon-yellow flowers are up to 3cm in
diameter. The flat fruits have papery lobes.
The fleshy roots are slimy when eaten,
although they are enjoyed by duikers.

   Found on sandy flats, often at the foot
of hills, in Namaqualand, the Hantam
Mountains and from the Tankwa Karoo
to Robertson.
Flowers July to October.

RUMEX
There are 17 species of *Rumex* (14 indigenous,
three alien) in southern Africa of which three (two
indigenous, one alien) occur in Namaqualand.

Rumex cordatus
rooisuring
POLYGONACEAE

A deciduous perennial herb with erect
branches up to 50cm high, with a tuber.
The basal leaves, up to 9cm long, are
egg-shaped and have rounded basal lobes.
Reddish flowers are borne on an erect
stalk. Fruits are reddish and winged.
    Found on sandy flats, often on old
fields, in Namaqualand, extending into
the Western and Eastern Cape.
Flowers July to September.

ZELDA WAHL

DROSERA
There are 20 species of *Drosera* in southern Africa of
which two occur in Namaqualand.

Drosera trinervia
sundew, sonnedou, snotrosie
DROSERACEAE

A deciduous perennial herb up to 10cm
high. The inversely sword-shaped leaves,
clustered close to the ground, have very
long, reddish glandular hairs on the surface.
Red, white or mauve flowers are borne on
leafless stalks up to 10cm long; the stigmas
are divided into many segments.
    Found in seasonally very damp sand in
the Namakwaland Klipkoppe and in the
Western Cape.
Flowers August to October.

PIETER VAN WYK

LITA COLE

FRANKENIA
There are four species of *Frankenia* in southern Africa
of which three occur in Namaqualand.

Frankenia repens
sea heath, soutrankie
FRANKENIACEAE

A creeping, mat-forming shrub up to 35cm
in diameter. The slightly fleshy, dark grey-
green leaves are cylindrical. Pink to mauve
flowers are borne at the ends of branches.
　　Found in very saline sandy soil, usually
on or next to pans or river mouths, in
the Namakwaland Klipkoppe and on the
Coastal Plain, extending into the Western
and Eastern Cape.
Flowers September to October.

NORBERT JÜRGENS

TAMARIX
There are five species of *Tamarix* (one indigenous, four
alien) in southern Africa of which two (one indigenous,
one alien) occur in Namaqualand. *Tamarix usneoides* is
the only indigenous species in southern Africa.

Tamarix usneoides
abikwa tree, dabbeboom
TAMARICACEAE

A tree or large shrub up to 5m tall, with
hanging branches. Bark brownish-grey
with rough transverse scars. Small, pale
grey-green leaves overlap each other. Many
small, scented, white or cream-coloured
flowers borne close together on a long
stalk; anthers pink; male and female
flowers are borne on separate plants.
　　Found in dry sandy areas in and
around pans and riverbanks in the
Richtersveld and Namakwaland Klipkoppe,
also in other Karoo areas and in Namibia.
Flowers January to December after rain.

Viscum capense
voëlent

VISCACEAE

A green to blue-green, aerial hemiparasite
growing in dense rounded clumps on
shrubs or trees. The rounded branches
have only scale-like leaf rudiments. Yellow
or cream-coloured flowers are borne
singly on the nodes. The round fruits are
white, almost translucent, and up to 4mm
in diameter.

Found on trees and shrubs along dry
watercourses throughout Namaqualand
and in the rest of southern Africa.
Flowers June to November.

ZELDA WAHL

Viscum rotundifolium
voëlent

VISCACEAE

A green to blue-green, aerial hemiparasite
growing in rounded clumps on shrubs
or trees. The young green branches are
ribbed. The slightly fleshy, elliptic leaves
are up to 1.5cm long and 1cm across. Small,
yellow flowers are borne in small clusters
in leaf axils; male and female flowers are
borne on the same stalk. The round, orange
fruits are up to 5mm in diameter.

Found on trees and shrubs in Namaqualand and in the rest of southern Africa.
Flowers February to May.

ANNELISE LE ROUX

VISCACEAE **255**

ZELDA WAHL

There are about 170 species of *Thesium* in southern Africa of which 12 occur in Namaqualand.

Thesium lineatum
SANTALACEAE

An erect, blue-green, woody perennial shrub up to 1m high, with grooved branches. The few inconspicuous leaves are linear, flattened and slightly fleshy. Small, white flowers are borne singly in axils of already shed leaves. The round to oblong fruits are white when mature.

Found on rocky slopes in Namaqualand, extending southwards to the Swartberg and northwards into Namibia. **Flowers March to October.**

HEATHER BURGER

MOQUINIELLA
There is one species of *Moquiniella* in southern Africa and in Namaqualand.

Moquiniella rubra
kootjie-nam-nam
LORANTHACEAE

An aerial hemiparasite growing on shrubs and trees. The elliptic leaves are up to 4cm long and 1cm across. Red flowers, tubular with a swollen base, are borne 3–6 together in leaf axils; the 5 lobes coil back; the style is much longer than the flower tube. The round fruits are red when mature.

Found on trees and large shrubs in the Richtersveld and Namakwaland Klipkoppe, extending into the Western and Eastern Cape. **Flowers March to September.**

Septulina glauca
kootjie-nam-nam, vuurhoutjies

LORANTHACEAE

An aerial hemiparasite growing on shrubs or trees; its greyish branches bear raised leaf scars. The slightly fleshy, grey, broadly elliptic leaves are covered with a powdery down. Red flowers, tubular and swollen and paler at the tip, are borne in leaf axils; the 4 lobes bend backwards; the style is just longer than the flower tube. The round to oblong fruits are red when mature.

Found on trees in Namaqualand, extending southwards into the Western Cape and northwards into Namibia.
Flowers March to November.

PIETER VAN WYK

TAPINANTHUS
There are five species of *Tapinanthus* in southern Africa of which one occurs in Namaqualand.

Tapinanthus oleifolius
nam-nambos

LORANTHACEAE

An aerial hemiparasite growing on shrubs or trees. The yellow-green, broadly elliptic leaves are up to 4.5cm long and 2cm across. Red flowers, tubular and with a conspicuous swollen base, are borne 3 or 4 together in leaf axils; the 5 lobes bend backwards; the style is just longer than the flower tube. The elliptic fruits are reddish-orange when mature.

Found on trees in Namaqualand and in other dry areas of Africa south of the equator.
Flowers January to December.

VERONICA ESTEHUYSEN

LORANTHACEAE **257**

ZELDA WAHL

There is only one species of *Dyerophytum* in southern Africa and in Namaqualand.

Dyerophytum africanum
PLUMBAGINACEAE

A laxly growing shrub up to 50cm high. The broadly spoon-shaped leaves, up to 2cm long and 1cm across, are often coated with calcareous granules. Flowers are borne densely along the ends of branches; the inconspicuous funnel-shaped, yellowish flower tube is hidden in a 5-winged, transversely wrinkled, papery, pinkish calyx that is persistent for a long time. A palatable plant to stock.

Found on rocky slopes in the Richtersveld and Namakwaland Klipkoppe, extending eastwards into Bushmanland and northwards through Namibia into Angola.
Flowers March to October.

ANNELISE LE ROUX

LIMONIUM
There are about 20 species of *Limonium* in southern Africa of which eight occur in Namaqualand. Two species are still undescribed.

Limonium namaquanum
PLUMBAGINACEAE   endemic

A gnarled, evergreen shrub up to 30cm high, with a woody base easily splitting into strips. The slightly fleshy, greyish, spoon-shaped leaves, up to 3.5cm long, are arranged in 2 opposite rows on the branch. Pink flowers are closely set on round stalks at the ends of branches; the papery, showy pink to cream-coloured calyx looks like a flower and remains persistent for a long time after the flower has died.

Found among quartz pebbles on gentle slopes on the Coastal Plain.
Flowers August to December.

Limonium sinuatum
statice, papierblommetjie

PLUMBAGINACEAE   alien invader

A very hairy, deciduous perennial herb up to 40cm high, with a woody root. The basal leaves are deeply lobed and about 20cm long. White or pink flowers are closely set at the ends of winged stalks; the papery, showy white, pale yellow, mauve, pink or purple calyx looks like a flower and remains persistent for a long time after the flower has died. Introduced from the Mediterranean region and has escaped from cultivation.

Found mostly in sandy, disturbed soils in Namaqualand, extending into the Western and Eastern Cape.
Flowers June to November.

ANNELISE LE ROUX

Limonium teretifolium
(formerly *Afrolimon teretifolium*)

PLUMBAGINACEAE   endemic

A small, erect shrub up to 20cm high. The fleshy, deep green leaves are hard and cylindrical. Pink flowers are closely set on round stalks at the ends of branches; the papery, showy white calyx looks like a flower and remains persistent for a long time after the flower has died.

Found among quartz pebbles on the Knersvlakte.
Flowers June to October.

ANNELISE LE ROUX

PLUMBAGINACEAE **259**

SPERGULARIA
There are five species of *Spergularia* in southern
Africa of which three occur in Namaqualand.

Spergularia bocconii
CARYOPHYLLACEAE

A spreading, slender annual or biennial
herb up to 25cm high, with glandular
hairs on the branches that are thickened
at the nodes. The fleshy, linear leaves are
in opposite pairs, with each pair at right
angles to the one above and below it;
large, membranous stipules encircle the
base of the leaf pair. The white flowers
have up to 10 stamens.

Found in sandy soils on the Coastal
Plain, extending into the Western and
Eastern Cape.
Flowers May to September.

SILENE
There are about 15 species of *Silene* in southern Africa
of which four occur in Namaqualand.

Silene undulata
Job-se-kraaltjies
CARYOPHYLLACEAE

An erect or sometimes sprawling,
deciduous perennial herb up to 60cm
high. The basal, spoon-shaped leaves are
up to 5cm long and 1.5cm across. White,
lemon or pink flowers are borne on a
loosely branched stalk up to 50cm long.

Found on sandy flats in Namaqualand
and in the rest of southern Africa.
Flowers August to October.

DIANTHUS
There are about 20 species of *Dianthus* in southern
Africa of which three occur in Namaqualand.

Dianthus kamisbergensis
Kamiesberg carnation, Kamiesberg-
angeliertjie

CARYOPHYLLACEAE   endemic

A perennial herb growing in a grass-like
tuft up to 30cm high. The linear, grey-
green leaves gradually taper towards the
pointed tip; the leaf surface is covered
with very short, stiff hairs towards the
base. Dark pink flowers about 3cm in
diameter are borne on a loosely branched
stalk; the petals are spreading, with a
shallowly toothed tip.

Found in sandy soil on the Kamiesberg.
Flowers November.

ANNELISE LE ROUX

Dianthus namaensis
grass carnation, grasangeliertjie

CARYOPHYLLACEAE

A perennial herb growing in a grass-like
tuft up to 30cm high. The linear, blue-
green leaves, tinged purple with age,
are up to 6cm long and 2mm across and
taper gradually towards the pointed tip.
White or pale pink flowers about 2.5cm in
diameter are borne on a loosely branched
stalk; the petals are spreading, with a
deeply dissected fringe.

Found on rocky slopes in Namaqualand,
extending into Bushmanland, the Western
and Eastern Cape and into Namibia.
Flowers August to December.

ANNELISE LE ROUX

CARYOPHYLLACEAE **261**

ZELDA WAHL

There are about 15 species of *Hermbstaedtia* in southern Africa of which one occurs in Namaqualand.

Hermbstaedtia glauca
bokhout
AMARANTHACEAE

An erect shrub up to 70cm high, with blue-green branches. The erect, blue-green leaves, up to 2.5cm long and 2mm across, are linear-oblong. Mauve-pink to cream-coloured flowers are borne in a dense, round cluster at the ends of branches.

Found in sandy soil, often in dry riverbeds in the Richtersveld and Namakwaland Klipkoppe, extending eastwards into Bushmanland and northwards into southern Namibia. **Flowers July to October.**

NORTBERT JÜRGENS

CALICOREMA
There are two species of *Calicorema* in southern Africa of which one occurs in Namaqualand.

Calicorema capitata
AMARANTHACEAE

A shrub up to 75cm high, with grey-green branches. The short-lived, linear leaves are up to 2cm long and 1.5mm across. Flowers covered with dense, long, white, silky hairs are borne at the ends of branches; the stamens are dark pink.

Found in sandy soil and in dry riverbeds in the Richtersveld, extending eastwards into Bushmanland and northwards into southern Namibia. **Flowers September to October.**

CAROXYLON
There are about 85 species of *Caroxylon* in southern Africa of which 14 occur in Namaqualand.

Caroxylon aphyllum
(formerly *Salsola aphylla*)
blomkoolganna
AMARANTHACEAE

A much-branched shrub up to 1m high. The slightly fleshy, scale-like, grey-green leaves are closely arranged on the branch; the leaf surface is hairy, soon becoming hairless and wrinkled. The solitary flowers are minute and inconspicuous. The papery, broadly winged fruits are green to light brown.

Found on sandy flats in Namaqualand, extending into the Western Cape, Eastern Cape and Namibia.
Flowers December to February.

TESSA OLIVER

KALI
There is one alien species of *Kali* in southern Africa and in Namaqualand.

Kali sp. (formerly *Salsola kali*)
tumbleweed, rolbos
AMARANTHACEAE   alien

A rigid, spiny annual herb up to 50cm high, with fluted branches. The triangular leaves, up to 1.2cm long, are spine-tipped. The papery fruits are hard, brown or brownish-yellow. When the plant is dry, it is easily dislodged and tumbled around by the wind, a very effective form of seed dispersal. A weed introduced from Asia. An unpalatable plant except in the summer months when still young and green.

Found in sandy soil in disturbed areas in Namaqualand, extending into the Northern Cape, Western Cape and Namibia.
Flowers September to November.

ZELDA WAHL

ANNELISE LE ROUX

SARCOCORNIA
There are about 10 species of *Sarcocornia* in southern
Africa of which six occur in Namaqualand.

Sarcocornia littorea
sea coral, seekoraal
AMARANTHACEAE

A spreading shrub up to 50cm high. The
very fleshy, bright green, barrel-shaped or
cylindrical segments on the branches are
up to 8mm long and 5mm in diameter.
Only the stamens and stigmas of the
flowers show at the nodes.

Found among rocks in the sea, visible
only at low tide on the Coastal Plain,
extending to Port Elizabeth.
Flowers January to December.

NORTBERT JÜRGENSS

Sarcocornia natalensis
var. *affinis*
marsh coral, vleikoraal
AMARANTHACEAE

An erect shrub up to 20cm high, with
slightly woody, upright branches, the
base often bent into an S-shape; the
branches sometimes root at the nodes.
The very fleshy, cylindrical segments on
the branches are up to 3cm long and
5mm in diameter; the leaves are bright
green when the roots of the plant are
still in the water, but turn brown to red
as the environment dries out. Only the
stamens and stigmas of the flowers show
at the nodes.

Found in salt marshes in coastal
lagoons and inland salt pans on the
Coastal Plain, extending from Angola
along the coast to Mozambique.
Flowers September to May.

Sarcocornia pillansii
coral bush, litjiesbos
AMARANTHACEAE

A much-branched shrub up to 70cm high.
The fleshy, reddish-purple, cylindrical
segments on the branches are up to 2cm
long and 5mm in diameter; the corky dead
leaves adhere to the branches but older
branches become bare with time. Only the
stamens and stigmas of the flowers show
at the nodes.

Found on the edge of estuaries and
saline, inland marshes, indicating the
upper floodplain, in Namaqualand,
extending into the Western Cape, Eastern
Cape and Namibia.
Flowers January to June.

ANNELISE LE ROUX

Sarcocornia xerophila
litjiesbos
AMARANTHACEAE   endemic

A much-branched shrub up to 50cm high.
The greyish-green or bluish-green, barrel-
shaped or cylindrical segments on the
branches are up to 7mm long and 6mm in
diameter; the leaves change to a maroon
colour in summer.

Found among white quartz pebbles on
the Knersvlakte.
Flowers September to December.

ANNELISE LE ROUX

ANNELISE LE ROUX

MANOCHLAMYS
There is only one species of *Manochlamys* in southern
Africa and in Namaqualand.

Manochlamys albicans
spanspekbos, seepbos, bobbejaanseep
AMARANTHACEAE

A loosely branched shrub up to about 70cm
high, with pale-coloured branches. The grey,
arrowhead-shaped leaves are up to 4cm
long and 3.2cm across. Minute yellowish
flowers are crowded near the ends of
branches. The fleshy, green to yellow fruits
smell like melon. The fruits can be used like
soap as they foam when rubbed between
the hands. Very palatable, mostly grazed in
the summer months when other palatable
plants have shed their leaves.

Found on sandy flats and rocky slopes
in Namaqualand, extending southwards
into the Western Cape and northwards
into southern Namibia.
Flowers August to January.

ZELDA WAHL

ATRIPLEX
There are 15 (10 indigenous, five alien) species of
Atriplex in southern Africa of which eight (five
indigenous, three alien) occur in Namaqualand.

Atriplex lindleyi
subsp. *inflata*
klappiesbrak, blasiebrak
AMARANTHACEAE   alien

An erect shrub up to 40cm high; branches
whitish; separate male and female plants.
Greyish, triangular leaves up to 3cm long
and 2.5cm across; margin slightly toothed.
Inconspicuous flowers borne in clusters
in leaf axils. Fruits winged and spongy.
Introduced from Australia, this unpalatable
plant is now 'naturalised' and a pioneer
plant, being among the first to colonise
disturbed areas. It is browsed only in the
summer months when food is very scarce.

Found on sandy, rocky flats, especially in
disturbed areas throughout southern Africa.
Flowers May to October.

Atriplex semibaccata
creeping saltbush, rankbrakbossie
AMARANTHACEAE

A sprawling annual or perennial herb up to 25cm high, with separate male and female plants. The greyish, broadly elliptic leaves are up to 2cm long and 5mm across; the leaf surface is irregularly toothed. Inconspicuous flowers are borne in clusters in leaf axils. The fleshy, red fruits are round to elliptic.

Found on sandy flats, often in disturbed areas, in Namaqualand and also in the rest of southern Africa.
Flowers September to December.

ANNELISE LE ROUX

Atriplex vestita
sea saltbush, seebrakbos
AMARANTHACEAE

An erect shrub up to 1.5m high, with separate male and female plants. The pale-coloured branches are angled when young, peeling later. The oval to elliptic leaves are borne on grey stalks that become slightly purple with age; the margin is smooth or has spreading lateral, blunt teeth at the leaf base. The flowers are pale yellow. The flat, oval to elliptic or slightly 3-lobed fruits are warty; the margin has regular, small teeth.

Found on the dunes closest to the sea along the coast of the Coastal Plain and in the Western and Eastern Cape.
Flowers August to December.

ZELDA WAHL

AMARANTHACEAE **267**

ANNELISE LE ROUX

LIMEUM
There are about 15 species of *Limeum* in southern
Africa of which four occur in Namaqualand.

Limeum africanum
koggelmandervoet
LIMEACEAE

A prostrate, branched annual or perennial
herb up to 35cm in diameter. The slightly
fleshy, bluish-green, linear to inversely
sword-shaped leaves are up to 3cm long;
the margin is sometimes rolled backwards.
White flowers are borne in dense, round
clusters at the ends of branches; the
petals are smaller than the calyx, which is
white with a broad green keel. A highly
palatable plant.

Found on sandy flats in Namaqua-
land, extending southwards to the Cape
Peninsula.
Flowers May to November.

ZELDA WAHL

Limeum fenestratum
LIMEACEAE

A slender, erect annual herb up to 50cm
high. The linear or lance-shaped leaves, up
to 5cm long, are sharply pointed or round-
tipped. Solitary, stalkless, white flowers
are borne in the forks of the branches.
Fruits have a flat, round, membranous,
transparent wing and is pale yellow, with
radiating green nerves.

Found in deep, red or yellow-brown
sand on the Coastal Plain and in the rest
of southern Africa.
Flowers June to April.

ADENOGRAMMA
There are about 10 species of *Adenogramma* in
southern Africa of which four occur in Namaqualand.

Adenogramma glomerata
muggiegras
MOLLUGINACEAE

A prostrate, much-branched annual herb
up to 30cm in diameter, with brown to
orange branches. The linear leaves, borne
in whorls at intervals along the branches,
are up to 1cm long and tipped with small,
sharp, rigid hairs. White flowers about
3mm in diameter are borne in small
clusters in leaf axils. A very palatable and
sought-after plant.

Found in sandy flats in Namaqualand,
extending southeastwards to Humansdorp
and northwards into Namibia.
Flowers August to October.

ZELDA WAHL

HYPERTELIS
There are seven species of *Hypertelis* in southern
Africa of which three occur in Namaqualand.

Hypertelis angrae-pequenae
MOLLUGINACEAE

A spreading, densely branched shrub up
to 20cm high. The crowded, fleshy, grey-
green leaves, up to 1cm long, are oblong-
spoon-shaped. Flowers are borne on stalks
about 2.5cm long; the petals are white,
with a green median stripe.

Found on sandy flats on the Coastal
Plain, extending northwards into Namibia.
Flowers May to October.

ANNELISE LE ROUX

MOLLUGINACEAE **269**

ZELDA WAHL

Hypertelis salsoloides
haassuring
MOLLUGINACEAE

A much-branched shrub up to 25cm high. The fleshy, cylindrical, blue-green leaves are up to 3.5cm long. White to pink flowers about 8mm in diameter are borne in sparse clusters at the ends of warty, leafless stalks up to 15cm long; the petals bend backwards. The leaves of this plant taste sour, almost like the leaves of *Oxalis*.

Found in sandy soil in Namaqualand and in the rest of southern Africa. **Flowers July to October.**

MARYNA KOHRS

Hypertelis spergulacea
MOLLUGINACEAE

A shrub up to 25cm high. The blue-green leaves, borne in whorls of 6 or 7 along the branches separated by long, slender internodes, are up to 2cm long. White to pink flowers are borne in clusters of 3–6 at the end of a slender stalk up to 4cm long.

Found in sandy soil, often in dry watercourses in the Richtersveld, extending eastwards into Bushmanland and northwards into Namibia. **Flowers June to September.**

PHARNACEUM
There are about 25 species of *Pharnaceum* in southern
Africa of which 11 occur in Namaqualand.

Pharnaceum croceum
MOLLUGINACEAE

An annual herb up to 15cm high. The
linear leaves, in tufted basal rosettes,
are up to 4cm long. White or pale pink
flowers about 7mm in diameter are borne
loosely on branched, almost leafless stalks;
the petals bend backwards; the 5 stamens
bear bright orange to red anthers. The
flowers open only in the late afternoon.

Found on sandy flats in Namaqualand,
extending southeastwards to Calvinia, on
to the Roggeveld Escarpment, to George
and northwards into Namibia.
Flowers May to October.

ZELDA WAHL

Pharnaceum lanatum
wilde-anys
MOLLUGINACEAE

An erect shrub up to 40cm high. The
narrow, linear leaves, up to 3cm long,
are evenly spaced along the branch; the
margin is rolled from both sides to the
middle of the lower side and tipped with
a hair-like point; the basal appendage on
the leaf stalk is white, with many woolly
hairs forming a loose or dense entangled
tuft. White flowers with a median green
stripe are borne in loose clusters at the
ends of leafless stalks. The flowers open
only in the late afternoon.

Found on rocky flats and slopes in
Namaqualand, extending southwards to
the Cape Peninsula.
Flowers August to October.

ZELDA WAHL

MARYNA KOHRS

PORTULACARIA
There are seven species of *Portulacaria* in southern
Africa of which four occur in Namaqualand.

Portulacaria fruticulosa
dwarf porkbush, suikerkanna,
dwerg-wolftoon
DIDIERIACEAE

An erect, much-branched shrub up to
70cm high, with slender, reddish-brown
branches. The fleshy, flattened leaves are
up to 6mm long. Pink flowers are borne
close together at the ends of branches.
The deep red fruits are hemispherical and
up to 1cm long and 6mm in diameter.

Found on rocky flats and slopes in
the Richtersveld, extending eastwards
into Bushmanland and northwards into
southern Namibia.
Flowers November to May.

Portulacaria namaquensis
Namaqua porkbush, wolftoon
DIDIERIACEAE

An erect, much-branched shrub up to
2m high, with smooth, white branches.
The fleshy, ovoid, blue-green leaves, up
to 4mm long, are borne in small clusters
along the branches. Small, pink flowers
are borne on sparse stalks at the ends of
branches. Fruits are roundish.

Found on rocky slopes in the
Richtersveld, extending eastwards into
Bushmanland and northwards into
southern Namibia.
Flowers December to February.

ANNELISE LE ROUX

ANACAMPSEROS
There are about 30 species of *Anacampseros* in
southern Africa of which 19 occur in Namaqualand.

Anacampseros herreana

ANACAMPSEROTACEAE   endemic

A perennial succulent up to 5cm high,
with decumbent, slender branches. The
leaves are partly covered by white, papery
scales that are broadly egg- to sword-
shaped and slightly hooded. White flowers
up to 1cm in diameter are borne singly
at the ends of branches; the petals are
almost as long as broad; the style is longer
than the stamens.

Found among rocks, often among
quartz pebbles, on slopes in the
Richtersveld.
Flowers October to December.

ANNELISE LE ROUX

Anacampseros lanceolata

ANACAMPSEROTACEAE

A perennial succulent up to 10cm high.
The very fleshy, sword-shaped leaves,
up to 2cm long and broader than thick,
are blue-green at the beginning of the
growing season, but become wrinkled and
brown at flowering time; the long, white
hairs in the leaf axils are often longer than
the leaves. Pink or white flowers more
than 2cm in diameter are borne in a small
cluster on a central fleshy stalk with long
hairs on the nodes; the petals are almost
as broad as long; the style is longer than
the stamens.

Found on rocky flats or slopes in
Namaqualand and in the Little Karoo.
Flowers October to November.

ADRIAN FORTUIN

Anacampseros namaquensis
hasiekos
ANACAMPSEROTACEAE

An erect perennial succulent up to 10cm high, branched at ground level. The fleshy, tightly packed leaves are as broad as they are thick, and are rounded to flat on top; long, white, silky hairs are borne in the leaf axils. Pink flowers up to 2cm in diameter are borne in a small cluster on a stalk with short hairs at the nodes; the petals are twice as long as broad; the style is longer than the stamens.

Found in cracks on flat rocks in Namaqualand and the Tankwa Karoo. **Flowers October to December.**

Anacampseros papyracea
gansmis
ANACAMPSEROTACEAE

An erect perennial succulent up to 5cm high, branched at ground level. The leaves are minute and hidden by white, papery, overlapping scales that are broadly egg- to tongue-shaped, with a serrate to toothed margin. White to pinkish flowers up to 1cm in diameter are borne singly at the ends of branches; the petals are slightly longer than broad; the style is shorter than the stamens.

Found among rocks, often in quartz, on flats in the Richtersveld and Namakwaland Klipkoppe, extending from Namibia to the southern Karoo. **Flowers October to December.**

Anacampseros quinaria
moerplant
ANACAMPSEROTACEAE

A perennial dwarf succulent, much-branched from a flat-topped, underground stem; the vegetative branches are less than 1cm high, while the flower-bearing branches are up to 3cm high. The leaves are minute and hidden by white, triangular, overlapping scales, darkening towards the tip. White or pink flowers up to 2cm in diameter are borne singly at the ends of branches; the petals are slightly longer than broad.

Found on rocky flats in the Namakwaland Klipkoppe, extending eastwards into Bushmanland and northeastwards into Namibia.
Flowers October to January.

EVA BRAND

Anacampseros vanthielii
ANACAMPSEROTACEAE   endemic

A clump-forming perennial succulent up to 2cm high. The fleshy, tightly packed leaves are as broad as they are thick, and are flat and triangular at the top. Pink flowers up to 4cm in diameter are borne in a small cluster on a fleshy, down-curving stalk with long, white, silky hairs at the base between the leaves; the petals are as long as broad; the style is longer than the stamens.

Found in cracks of rocks in the Namawaland Klipkoppe.
Flowers in September.

JEAN SMITH

FRUITS OF THE VYGIE GROUP IN THE FAMILY AIZOACEAE
Note: Numerals in brackets represent number of valves

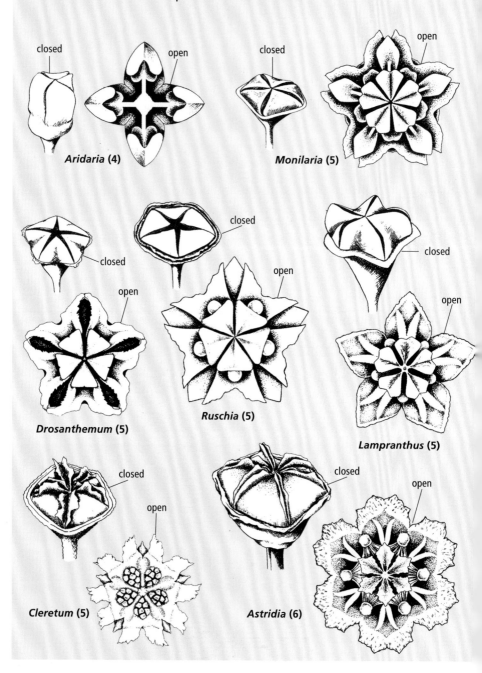

Aridaria (4)

Monilaria (5)

Drosanthemum (5)

Ruschia (5)

Lampranthus (5)

Cleretum (5)

Astridia (6)

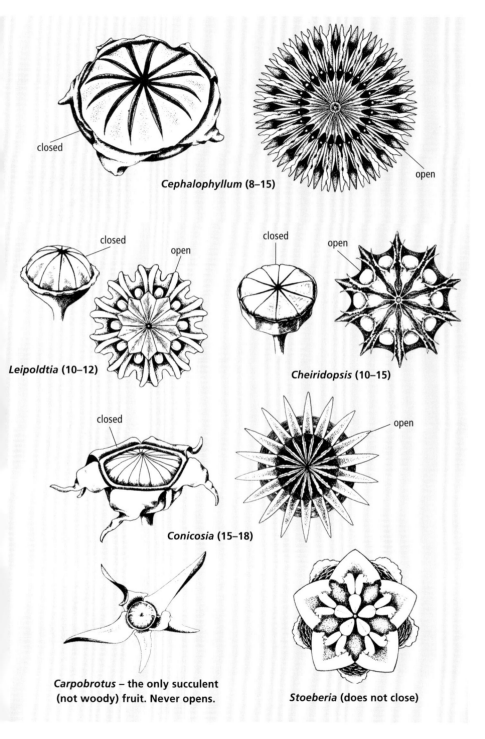

closed

Cephalophyllum (8–15)

open

closed

open

Leipoldtia (10–12)

closed

open

Cheiridopsis (10–15)

closed

open

Conicosia (15–18)

Carpobrotus – the only succulent (not woody) fruit. Never opens.

Stoeberia (does not close)

PIETER VAN WYK

Sesuvium sesuvioides
AIZOACEAE

A prostrate perennial herb; the pale-coloured branches, up to 50cm long, have coarse, white papillae. The slightly fleshy, elliptic-oblong leaves, up to 10cm long and 1cm across, are folded on the midrib. The petals of the magenta-pink flowers are fused for a third to half of their length and are sharply pointed; there are 5–10 stamens.

Found on sandy flats or among rocks in the Richtersveld and elsewhere in southern Africa, tropical Africa and India. **Flowers August to October.**

MARYNA KOHRS

Trianthema parvifolia
AIZOACEAE

An annual or sometimes perennial herb with smooth, yellow-brown or red, robust branches lying flat on the ground and spreading up to 50cm in diameter. The fleshy, green, often reddish, stalked leaves, up to 12mm long, are oblong to egg-shaped to almost round. Small, pinkish-purple to purple-red flowers are borne in clusters; there are 5 stamens.

Found in reddish, sandy loam soil on flats in the Richtersveld, extending into the Western Cape and Namibia. **Flowers July to September.**

AIZOON
There are 12 species of *Aizoon* in southern Africa of
which three occur in Namaqualand.

Aizoon canariense
galsiekslaai
AIZOACEAE

A prostrate annual herb with slightly
fleshy branches. The slightly fleshy, oblong
to inversely egg-shaped leaves are up to
5cm long and 1.5cm across. Dull yellow-
green flowers about 3mm in diameter are
borne directly on the branches. Fruits are
fleshy at first, becoming woody later and
are also borne directly on the branches.

Found in sandy soil in the Richtersveld
and Namakwaland Klipkoppe, also
in other dry regions of South Africa,
Namibia, northern Africa and Arabia.
Flowers May to December.

HEATHER BURGER

TETRAGONIA
There are about 35 species of *Tetragonia* in southern
Africa of which 14 occur in Namaqualand.

Tetragonia decumbens
AIZOACEAE

A sprawling shrub up to 45cm high,
with branches up to 1m long. The fleshy,
inversely egg-shaped to oblong leaves, flat
or more commonly recurved at the edges,
are covered with hairs. Yellow flowers,
smaller than the leaves, are borne in
clusters of 3–5; stamens are numerous. The
4-winged fruit is flat on top, with a small
intermediate ridge between the wings.

Found on white sand immediately
above the high-water mark on the
Namaqualand coast, extending from
Namibia to the Eastern Cape.
Flowers August to December.

ANNELISE LE ROUX

AIZOACEAE **279**

Tetragonia echinata
(previously misidentified as *Tetragonia microptera*)
misbredie
AIZOACEAE

A prostrate annual herb with slightly fleshy branches up to 30cm long. The stalked leaves, up to 2cm long and 1.5cm across, are broadly egg-shaped or round. Greenish-yellow flowers about 2mm in diameter are borne in clusters of 2–5. Fruits are round, with spiny ridges and horns. Grazed by sheep and goats.

Found in sandy flats, often in disturbed areas, in Namaqualand, extending from Namibia to the Western and Eastern Cape. **Flowers August to October.**

ZELDA WAHL

Tetragonia fruticosa
slaaibos, waterslaaibos
AIZOACEAE

An erect shrub up to 75cm high. The slightly fleshy leaves, up to 4cm long, are sword-shaped to oblong, tapering towards the base but without leaf stalks and flat or rolled; the midrib is prominent on the underside; there are numerous glistening water cells on the leaf surface. Yellow flowers, about 5mm in diameter and often tinged red, are borne in clusters of 2–5 towards the ends of branches. Fruits are rounded at the base, flat on top and are strongly 4-winged, with a distinct small ridge on the upper half between the wings.

Found in sandy and rocky flats or slopes in Namaqualand and elsewhere in the Northern, Western and Eastern Cape. **Flowers July to October.**

ZELDA WAHL

Tetragonia nigrescens
AIZOACEAE

A prostrate perennial herb up to 10cm high, with a tuber. The stalked leaves are slightly fleshy, covered with protuberances and variable in size and shape, the larger lower leaves broadly oval or almost round and the upper smaller ones narrower and more pointed. Stalked yellow flowers, covered with small protuberances or hairs, are borne in clusters of 2–7. Fruits are 4-winged, indented at the top and slightly narrowed at the base; the wings are rather thin and smooth, with an inconspicuous tubercle between them.

Found in sandy or loamy soil in Namaqualand, extending into the Western and Eastern Cape.
Flowers June to October.

OLGA CRUZ

Tetragonia pillansii
AIZOACEAE   endemic

A much-branched perennial herb up to 30cm high. The slightly fleshy leaves, up to 3cm long and 8mm across, are oblong to sword-shaped; the margin curves backwards. Cream-coloured flowers, with up to 20 stamens, are borne along the branches. Fruits are round to cylindrical, ending abruptly as if cut straight across and have 8 longitudinal ridges.

Found in sandy soil on the Coastal Plain.
Flowers August to September.

ANNELISE LE ROUX

ZELDA WAHL

There are about 30 species of *Galenia* in southern Africa of which 16 occur in Namaqualand.

Galenia africana
kraalbos

AIZOACEAE

An erect, yellowish-green shrub up to 1.5m high. Linear leaves up to 5cm long and hairless or almost so. Inconspicuous yellowish-white flowers about 1.5mm in diameter borne in large, loose clusters at the ends of branches. This is the dominant plant throughout Namaqualand and not only an indicator of disturbance, but also a pioneer plant, being the first perennial shrub to colonise after disturbance. It can also be the only remaining species after the veld has been heavily over-grazed.

Found on sandy flats in Namaqualand and also widespread in dry areas of the Western Cape, Eastern Cape and Namibia. **Flowers September to October.**

HEATHER BURGER

Galenia fruticosa
vanwyksbossie

AIZOACEAE

A diffuse, rounded shrub up to 50cm high. The slightly fleshy, yellowish branches are thinly covered with hairs pressed against them. The oblong, grey or almost white, opposite leaves are densely covered with silky hairs and are usually folded upwards, with a sharply pointed, hooked tip. Pale to deep pink flowers with hairy petals are borne on a much-branched, recurved stalk; after flowering the branched stalk hardens and may become spiny.

Found in sandy soil on flats in Namaqualand, in the Great Karoo, the Little Karoo and in Namibia. **Flowers throughout the year but peaks September to November.**

Galenia meziana
platkraalbos
AIZOACEAE

A prostrate, spreading, yellow-green
shrub with branches up to 30cm long. The
yellow-green leaves, 5–20mm long, are
linear and recurved at the ends. Cream-
coloured to pink flowers up to 5mm in
diameter are borne in clusters at the ends
of branches.

Found in sandy soil in the Richtersveld,
Namakwaland Klipkoppe and on the
Coastal Plain, extending into Namibia.
Flowers July to November.

ZELDA WAHL

Galenia sarcophylla
beslyn, porselein, t'gouboedanna
AIZOACEAE

A prostrate, grey shrub up to 40cm long.
The slightly fleshy, greyish-green leaves,
up to 1.7cm long, are inversely sword-
shaped to oblong and densely covered
with minute grey hairs or large glistening
water cells. Yellow or pink flowers about
2mm in diameter are borne in clusters
along the branches.

Found throughout Namaqualand
and also throughout the dry areas of
the Northern, Western and Eastern
Cape and Namibia.
Flowers June to November.

ZELDA WAHL

AIZOACEAE **283**

ZELDA WAHL

Mesembryanthemum barklyi
olifantsoutslaai
AIZOACEAE

A biennial succulent up to 1m high when in flower. Branches 4-angled and winged. Plant first forms a basal rosette of leaves with flowering side branches, the internodes in the rosette later elongate to form additional rosettes and new flowering branches. Very large, fleshy, triangular, boat-shaped, blue-green leaves up to 40cm long and 25cm across; water-storing epidermal cells small and rounded. Flowers pink, white or slightly green; numerous staminodes and stamens; the flowers open from noon until dusk. Dry fruits soft, with 5 valves.

Found in sandy areas in the Richtersveld and on the Coastal Plain, into Namibia. **Flowers November to January.**

ANNELISE LE ROUX

Mesembryanthemum fastigiatum
AIZOACEAE   endemic

A tufted, erect herb up to 10cm high. The cylindrical leaves are up to 4cm long and 0.7cm across; the enlarged water-storing epidermal cells are usually conspicuous. The white flowers, with petals closely pressed together, are up to 1.5cm in diameter. The dry fruits are soft, with 5 valves.

Found in sandy soil, usually in disturbed areas on the Knersvlakte. **Flowers October to November.**

Mesembryanthemum guerichianum
soutslaai

AIZOACEAE

A prostrate annual succulent. The flat, egg-shaped leaves are up to 25cm long and 10cm across; the water-storing epidermal cells are very large and rounded to elongated; the margin is undulate. The white to pale pink flowers are up to 5.5cm in diameter; filamentous staminodes are present; the flowers open from noon until dusk. The dry fruits are soft, with 5 valves.

Found in sandy soil, usually in disturbed areas, in Namaqualand, extending into the Western Cape, Eastern Cape, Namibia and Angola.
Flowers September to December.

VERONICA ESTERHUIZEN

Mesembryanthemum hypertrophicum
kinderpieletjies

AIZOACEAE

A prostrate annual succulent up to 15cm high, with branches up to 50cm long. The fleshy, bright green, cylindrical leaves, up to 7cm long and 1cm in diameter, become brown with age; the water cells on the leaf surface are arranged in longitudinal rows. The white to pale yellow flowers, the upper parts sometimes pink, are up to 6cm in diameter; they are open day and night, emitting a scent at dusk. The dry fruits are soft, with 5 valves.

Found on the Coastal Plain and in Namibia.
Flowers November to January.

ANNELISE LE ROUX

UTE SCHMIEDEL

Mesembryanthemum nodiflorum
AIZOACEAE

A prostrate annual herb up to 3cm high. The fleshy, cylindrical leaves are up to 2cm long and 5mm in diameter; the water-storing epidermal cells are conspicuous and rounded. The white or pale pink flowers are up to 1cm in diameter. The dry fruits are soft, with 5 valves. Although this plant is an indigenous species in southern Africa, it is also found in northern Africa, the Middle East, Europe, Australia and the Canary Islands.

Found on sandy flats on the Coastal Plain, the Knersvlakte and also in other dry areas in the Western Cape and in Namibia. **Flowers September to November.**

ANNELISE LE ROUX

Mesembryanthemum amplectens
(formerly *Aspazoma amplectens*)

AIZOACEAE   endemic

A branched shrub up to 15cm high. Young branches green and fleshy; surface of the branches has closely packed, cylindrical water-storing epidermal cells. Fleshy, yellow-green leaves in opposite pairs with each at right angles to those above and below it; basally free but stem-clasping and enveloping each other; water-storing epidermal cells flattened. White to pale yellow flowers up to 4cm across, borne at the ends of branches; filamentous staminodes present. The seed-dispersal mechanism is a dry fruit with 4 or 5 valves that open when wet and close when the fruit dries.

Found on low quartz hills and gravelly plains in the Richtersveld and Namakwaland Klipkoppe. **Flowers September to October.**

Mesembryanthemum exalatum
(formerly *Sceletium exalatum*)
kougoed
AIZOACEAE

A prostrate shrub up to 15cm high. The slightly fleshy leaves, up to 3.5cm long and flat with a recurved tip, have a distinct, straight midvein with equally distinct curved lateral veins; the water-storing epidermal cells are small, but conspicuously rounded. The whitish flowers are about 2cm in diameter. As a seed-dispersal mechanism, the dry fruits have 4 valves that open when wet and close again when the fruit dries.

Found in sandy soil under bushes in the Namakwaland Klipkoppe, extending eastwards to Calvinia.
Flowers September to October.

VERONICA ESTERHUIZEN

Mesembryanthemum lilliputanum
(formerly *Phyllobolus abbreviatus*)

AIZOACEAE   endemic

A deciduous perennial succulent up to 5cm high, with a tuber. The fleshy leaves are almost cylindrical; the water-storing epidermal cells are large, rounded but sometimes elongated. The pale yellow flowers, with stamens and stigmas visible, are up to 2cm in diameter; filamentous staminodes are present. As a seed-dispersal mechanism, the dry fruits have 5 valves that open when wet and close again when the fruit dries.

Found in loamy soil among quartz pebbles on the Knersvlakte.
Flowers September to October.

TESSA OLIVER

UTE SCHMIEDEL

Mesembryanthemum nitidum
(formerly *Phyllobolus nitidus*)
AIZOACEAE

A spreading to cushion-shaped shrub up to 30cm high. The weakly lignified branches have a conspicuous cork layer. The fleshy, yellow-green leaves are almost cylindrical; the water-storing epidermal cells are conspicuous. The pale yellow flowers, sometimes with a reddish centre or salmon-pink tips, are up to 3cm in diameter; the stamens and stigmas are not concealed. As a seed-dispersal mechanism, the dry fruits have 4 or 5 valves that open when wet and close again when the fruit dries.

Found on flats in the Namakwaland Klipkoppe, on the Knersvlakte and also in other Karoo areas of the Western Cape. **Flowers August to October.**

ANNELISE LE ROUX

Mesembryanthemum prasinum
(formerly *Phyllobolus prasinus*)
AIZOACEAE   endemic

A deciduous, decumbent dwarf shrub up to 10cm high. Branches are weakly lignified, with a thick cork layer basally. The fleshy leaves are cylindrical to narrowly egg-shaped; the water-storing epidermal cells are large, sometimes flattened, sometimes elongated and tipped, but not hair-like. The greenish-yellow flowers are up to 4cm in diameter. As a seed-dispersal mechanism, the dry fruits have 5 valves that open when wet and close again when the fruit dries.

Found in loamy sand on plains or hills in the Richtersveld, Namakwaland Klipkoppe and on the Coastal Plain. **Flowers in August.**

Mesembryanthemum spinuliferum

(formerly *Phyllobolus spinuliferus*)

AIZOACEAE   endemic

An erect shrub up to 30cm high. Branches are very stout, fleshy and weakly lignified, but with a thick cork layer. The fleshy, almost cylindrical, narrowly egg-shaped leaves are in opposite pairs at right angles to the pair above and below it; the water-storing epidermal cells are small and flattened. The white, pale yellow to straw-coloured or pale salmon to dull orange flowers are up to 3cm in diameter; the stamens and stigmas are not concealed. As a seed-dispersal mechanism, the dry fruits have 4 valves that open when wet and close again when the fruit dries.

Found in red sandy soil on plains in Namaqualand.

Flowers September to October.

ANNELISE LE ROUX

Mesembryanthemum vanheerdei

(formerly *Phyllobolus roseus*)

AIZOACEAE

A densely branched shrub up to 1m high. The slightly fleshy, cylindrical leaves are persistent, becoming spiny when dry; the water-storing epidermal cells are fairly large. The pink to salmon flowers are up to 3cm in diameter; filamentous staminodes are present; the stamens and stigmas not concealed. As a seed-dispersal mechanism, the dry fruits have 5 valves that open when wet and close again when the fruit dries.

Found in sandy soil among granite boulders in the Namakwaland Klipkoppe.

Flowers September to October.

GRETEL VAN ROOYEN

ZELDA WAHL

Mesembryanthemum digitatum subsp. *digitatum*
duim-en-vinger

AIZOACEAE   endemic

A perennial succulent up to 15cm high, forming clumps with age. Extremely fleshy, finger-like leaves borne on a very constricted branch of only a millimetre or two; the angle between 'finger and thumb' is less than 90°; leaves wither completely into a white papery covering in the summer months. White flowers, with petals stiff and hard to the touch, are up to 2cm in diameter and remain continuously open for about three weeks; filamentous staminodes absent; stamens and stigmas are concealed. As a seed-dispersal mechanism, the dry fruits have 5 valves that open when wet and close again when the fruit dries.

Found among white quartz pebbles on the central and southern Knersvlakte. **Flowers October to December.**

ANNELISE LE ROUX

Mesembryanthemum digitatum subsp. *littlewoodii*
klein duimpie-en-vingertjie

AIZOACEAE   endemic

A perennial succulent up to 8cm high, forming clumps with age. The extremely fleshy, ovoid leaves are up to 5cm long; the angle between 'finger and thumb' is more than 90°; the leaves wither completely into a white papery covering in the summer months. Solitary, stalkless white flowers, with petals stiff and hard to the touch, are up to 1.5cm in diameter and remain continuously open for about three weeks; filamentous staminodes are absent; the stamens and stigmas are concealed. As a seed-dispersal mechanism, the dry fruits have 5 valves that open when wet and close again when the fruit dries.

Found among white quartz pebbles on the northern Knersvlakte. **Flowers October to December.**

Mesembryanthemum brevicarpum
(formerly *Aridaria brevicarpa*)
donkiebos
AIZOACEAE

An erect, branched shrub up to 1m high. The fleshy leaves are opposite, cylindrical, smooth and up to 3cm long. Flowers about 2.5cm in diameter are borne in candelabra form at the ends of branches; the petals are glistening white with light brown tips on the inside and yellow or light brown on the outside; the flowers are open during the day and close at dusk. As a seed-dispersal mechanism, the dry fruits have 4 or 5 valves that open when wet and close again when the fruit dries.

Found in Namaqualand, extending southwards to Clanwilliam and northwards into southern Namibia. **Flowers August to September.**

ANNELISE LE ROUX

Mesembryanthemum noctiflorum
(formerly *Aridaria noctiflora*)
vleisbos
AIZOACEAE

An erect shrub up to 1m high, with red-brown branches (the colour of raw meat, hence the vernacular name). Leaves fleshy, cylindrical, smooth, blue-green, up to 3cm long and 4mm in diameter. White flowers, suffused with various shades of pink, red, salmon, copper or yellow, borne at the ends of branches; flowers open at dusk and close in the morning. Fruit does not close completely again after its first opening and the central parts soon disintegrate, leaving an empty outer shell.

Found on rocky flats or slopes in Namaqualand, extending southeastwards into the Western and Eastern Cape and northwards into Namibia. **Flowers August to September.**

ANNELISE LE ROUX & PIETER VAN WYK

UTE SCHMIEDEL

Mesembryanthemum corallinum

(formerly *Brownanthus corallinus*)

coral mesem, koraalvygie

AIZOACEAE   endemic

An erect shrub up to 35cm high. Fleshy internodes on the branches usually rounded but sometimes cylindrical, resembling a string of beads; water-storing epidermal cells with vertical and horizontal dimensions almost equal. Fleshy, red-brown to purple-brown leaves are triangular in cross-section. White flowers up to 1.5cm in diameter are borne singly at the ends of branches. Dry fruits are up to 6mm in diameter; 5 valves open when wet and close again when the fruit dries.

Found among quartz pebbles on flats and slight slopes on the Coastal Plain and Knersvlakte, often gregarious and then giving a purplish tint to the vegetation. **Flowers in October.**

ANNELISE LE ROUX

Mesembryanthemum pseudoschlichtianum

(formerly *Brownanthus pseudoschlichtianus*)

AIZOACEAE

An erect shrub up to 70cm high and 1m in diameter. The fleshy, cylindrical internodes have brick-shaped water-storing epidermal cells, visible in cross-section as a thin white layer. The slightly fleshy, slightly triangular leaves with round tips have overlapping leaf bases, sometimes with an inconspicuous ring of hairs at the base. White flowers up to 1.5cm in diameter are borne in clusters at the ends of branches. The dry fruits are up to 5mm in diameter; as a seed-dispersal mechanism, the 4 valves open when wet and close again when the fruit dries.

Found on plains or foothills in the north of the Coastal Plain, extending into Namibia. **Flowers September to October.**

Mesembryanthemum tomentosum

(formerly *Brownanthus pubescens*)

AIZOACEAE

A cushion-shaped shrub up to 25cm high and 80cm in diameter. Internodes fleshy, cup-shaped with long, narrow water-storing epidermal cells. The slightly fleshy leaves are almost cylindrical; a conspicuous ring of white hairs at the base become united with age to form a white rim on the internode; water-storing epidermal cells are hair-like, giving a hairy appearance. Flowers up to 1.5cm in diameter borne singly at the ends of branches. Dry fruits are up to 6mm in diameter; as a seed-dispersal mechanism, the 5 valves open when wet and close again when the fruit dries.

 Found in rocky areas, often on quartz, in the Richtersveld, extending into southern Namibia.

Flowers August to September.

NORBERT JÜRGENS

Mesembryanthemum vaginatum

(formerly *Brownanthus ciliatus*)

AIZOACEAE

A decumbent to erect shrub up to 25cm high, 1m across. Fleshy branches with long, narrow water-storing epidermal cells. Pairs of slightly fleshy, almost cylindrical leaves are shortly fused towards the leaf bases; a ring of white hairs at the nodes on the branch covers less than half the internode length; water-storing epidermal cells are rounded or sometimes with pointed tips. White to cream-coloured flowers, up to 1.5cm across; filamentous staminodes absent. Dry fruits are up to 5mm across; as a seed-dispersal mechanism, the 5 valves open when wet, close when the fruit dries.

 On plains and along roadsides as a pioneer on the Knersvlakte, into the Tankwa Karoo, Great Karoo and Little Karoo.

Flowers October to December.

ANNELISE LE ROUX

ANNELISE LE ROUX

Mesembryanthemum dinteri
(formerly *Psilocaulon dinteri*)
kraalbossie
AIZOACEAE

A prostrate shrub up to 25cm high. Older branches distinctly jointed, fleshy, with barrel-shaped internodes if growing in saline coastal areas, otherwise cylindrical; branches change from blue-green to red at the end of the growing season; water-storing epidermal cells flattened or wart-like. Leaves slightly fleshy, cylindrical, upper surface somewhat flattened; leaves wither, dry and drop very early in the season. Pink, occasionally white flowers up to 1cm in diameter; filamentous staminodes gathered into a cone. As a seed-dispersal mechanism, the 5 valves open when wet and close again when the fruit dries.

Found in Namaqualand, extending into the Western Cape and Namibia. **Flowers September to January.**

ANNELISE LE ROUX

Mesembryanthemum leptarthron
(formerly *Psilocaulon leptarthron*)
AIZOACEAE   endemic

An erect to rounded shrub up to 1m high. The internodes on the green branches are distinctly jointed; the water-storing epidermal cells are flattened. The almost cylindrical or slightly triangular leaves have flattened water-storing epidermal cells. White flowers about 2.5cm in diameter are borne singly at the ends of branches; the filamentous staminodes are gathered into a cone. The dry fruits are 4.5–6mm in diameter; as a seed-dispersal mechanism, the 5 valves open when wet and close again when the fruit dries.

Found in firm, red, calcareous, loamy soil on plains, often common on 'heuweltjies' on the Knersvlakte. **Flowers August to October.**

Mesembryanthemum neofoliosum
(formerly *Psilocaulon foliosum*)

AIZOACEAE   endemic

A densely branched, spreading shrub up to 1m high and 1.5m in diameter. The almost cylindrical leaves are slightly flattened on the upper surface; the dry leaves remain on the branches; the water-storing epidermal cells are flattened. The white flowers are 1.5cm in diameter; the filamentous staminodes are gathered into a cone. The dry fruits are 3mm in diameter; as a seed-dispersal mechanism, the 5 valves open when wet and close again when the fruit dries.

Found in firm, reddish, loamy soil, often along roadsides, in the Namakwaland Klipkoppe and on the Coastal Plain.
Flowers September to November.

ANNELISE LE ROUX

Mesembryanthemum rapaceum
(formerly *Caulipsolon rapaceum*)

AIZOACEAE

A perennial herb with a tuber and creeping annual branches that are weakly lignified, green, smooth, fleshy and with conspicuous nodes. The fleshy, bright green, almost cylindrical leaves are smooth; the water-storing cells are flattened. The white flowers have a yellow centre and are up to 3cm in diameter; filamentous staminodes are absent. The dry, deeply funnel-shaped fruits are about 5mm in diameter; as a seed-dispersal mechanism, the 5 valves open when wet and close again when the fruit dries.

Found in red sand, often in disturbed areas, in Namaqualand, extending to Calvinia.
Flowers August to October.

ANNELISE LE ROUX

AIZOACEAE **295**

NORBERT JÜRGENS

Mesembryanthemum subnodosum

(formerly *Psilocaulon subnodosum*)
asbos
AIZOACEAE

An erect shrub up to 75cm high, with distinct joints on the green branches. The opposite young leaves have a short, stiff point at the end; the leaves are very short-lived and photosynthesis occurs mostly in the green branches. The white or pink flowers are up to 1.5cm in diameter; the filamentous staminodes are gathered into a cone. The dry fruits are up to 3.2mm in diameter; as a seed-dispersal mechanism, the 5 valves open when wet and close again when the fruit dries.

Found on sandy flats, usually in disturbed places, in Namaqualand, extending eastwards to Kenhardt.
Flowers August to December.

ANNELISE LE ROUX

Mesembryanthemum pallens

springbokslaai
AIZOACEAE   endemic

A deciduous, prostrate perennial shrub with branches up to 60cm long. The egg-shaped leaves are triangular in cross-section, with acute angles, and are slightly curved. The flowers are pink with a white centre or almost pure white; only a few filamentous staminodes are present; the filaments are white. As a seed-dispersal mechanism, the dry fruits have 4 or 5 valves that open when wet and close again when the fruit dries.

Found in sandy soil on flats and along roadsides in Namaqualand, extending southwards into the Western Cape.
Flowers September to October.

Mesembryanthemum sladenianum

(formerly *Prenia sladeniana*)

skotteloor

AIZOACEAE

A prostrate shrub with branches up to 50cm long. The 2 opposite, slightly fleshy, flat, almost circular young leaves adhere to each other along their margins, forming an almost globular structure inside which the following leaf pair develops, shielded against the harshness of the habitat; the leaves easily pop open when pressed. The flowers are white to slightly pink. As a seed-dispersal mechanism, the dry fruits have 4 valves that open when wet and close again when the fruit dries.

Found on rocky lower hills or in sandy riverbeds in the Richtersveld, extending into Namibia.

Flowers March to May.

ZELDA WAHL

Mesembryanthemum tetragonum

(formerly *Prenia tetragona*)

AIZOACEAE

A spreading to erect shrub up to 30cm high. The fleshy, blue-green leaves are almost cylindrical. White to pink flowers are borne on long pinkish stalks. As a seed-dispersal mechanism, the dry fruits have 4 valves that open when wet and close again when the fruit dries.

Found in sandy soil on plains in Namaqualand, extending into the Northern Cape, Western Cape, Free State and Namibia.

Flowers September to April.

NORBERT JÜRGENS

TESSA OLIVER

There are two species and three subspecies of *Conicosia* in southern Africa of which two species and one subspecies occur in Namaqualand.

Conicosia elongata
varkiesknol

AIZOACEAE

A deciduous perennial succulent with a large underground tuber. Leaves cylindrical, pointed, up to 11cm long and fleshy. White, cream-coloured or yellow flowers up to 6cm in diameter are borne singly on stalks. Dry conical fruits have 15–29 valves that break up into separate lightweight compartments when they are ready to disperse the seeds. The large, fleshy tuber is sought after by porcupines.

Found in flat sandy areas among hills, especially on abandoned cultivated fields, in Namaqualand, extending into the Western Cape up to Touwsrivier. **Flowers August to September.**

ZELDA WAHL

Conicosia pugioniformis subsp. *alborosea*
AIZOACEAE

A deciduous perennial succulent up to 40cm high, with a central rosette, side branches and a taproot. The fleshy, grey-green leaves are triangular in cross-section; not all the leaves are shed in the dry summer months. Pale yellow flowers up to 6cm in diameter are borne at the ends of side branches. The lower part of the fruit is cup-shaped and the conical upper part has 16–22 valves.

Found on sandy flats on the Coastal Plain, extending along the western and southern coast of the Western Cape. **Flowers September to November.**

There are 14 species of *Cleretum* in southern Africa of which seven species occur in Namaqualand.

Cleretum bruynsii
sandslaai

AIZOACEAE   endemic

A prostrate, rosette-forming annual herb up to 5cm high and 20cm in diameter. The slightly fleshy, flat, spoon-shaped, grey-green leaves are 30–50cm long and 6–14mm across; the water-storing epidermal cells are large and conspicuous. The yellow flowers are up to 1.5cm in diameter; filamentous staminodes are absent; stamens 7–10; stigmas 5. The dry fruits are 5–9mm in diameter; as a seed-dispersal mechanism, the 5 valves open when wet and close again when the fruit dries.

Found in deep, red sand in the Namakwaland Klipkoppe.
Flowers July.

HEATHER BURGER

Cleretum hestermalense
(formerly *Dorotheanthus bellidiformis* subsp. *hestermalensis*)
sandslaai

AIZOACEAE   endemic

A tufted annual herb up to 5cm high. The slightly fleshy, flattish, oblong leaves are up to 1.5cm long; the water-storing epidermal cells are large and glistening. The flowers, up to 3cm in diameter, are pale mauve with a white centre; the stamens are arranged in 1 or 2 rows. As a seed-dispersal mechanism, the dry fruits have 5 valves that open when wet and close again when the fruit dries.

Found in sandy soil in flat areas in the Namakwaland Klipkoppe and on the Coastal Plain.
Flowers August to September.

ANNELISE LE ROUX

Cleretum papulosum

(formerly *Cleretum papulosum* subsp. *papulosum*)
sandslaai
AIZOACEAE

A prostrate annual herb up to 3cm high. The slightly fleshy, flat, inversely sword-shaped leaves are up to 6cm long; the water-storing epidermal cells are not as large as those of *C. bruynsii*. The yellow flowers are up to 1.5cm in diameter; filamentous staminodes are absent. As a seed-dispersal mechanism, the dry, flat fruits have 5 valves that open when wet and close again when the fruit dries.

Found in sandy soil on flats in Namaqualand, extending to Mossel Bay in the Western Cape.
Flowers August to September.

ANNELISE LE ROUX

Cleretum patersonjonesii

sandslaai
AIZOACEAE   endemic

A prostrate annual herb up to 5cm high and 7–20cm in diameter. The fleshy, flat, spoon-shaped, grey-green leaves are up to 8cm long and 2cm across; the water-storing epidermal cells are large and conspicuous. The yellow flowers are up to 6cm in diameter; filamentous staminodes are absent; stamens numerous. The dry fruits are up to 1cm in diameter; as a seed-dispersal mechanism, the 5 valves open when wet and close again when the fruit dries.

Found in sandy soil in the fynbos or renosterbos vegetation of the Kamiesberg.
Flowers August to September.

COLIN PATERSON-JONES

Cleretum rourkei

(formerly *Dorotheanthus rourkei*)
sandslaai

AIZOACEAE   endemic

An annual herb up to 5cm high. The
fleshy, flat, inversely sword-shaped leaves
are up to 6cm long; the water-storing
epidermal cells are glistening. The red,
orange-red, salmon or yellow flowers
are about 3.5cm in diameter. As a seed-
dispersal mechanism, the dry, flat fruits
have 5 valves that open when wet and
close again when the fruit dries.

Found in deep, red, sandy soil on
plains from the Vanrhynsdorp vicinity to
Hondeklip Bay.
Flowers July to August.

ANNELISE LE ROUX

Cleretum schlechteri

(formerly *Cleretum papulosum* subsp. *schlechteri*)
sandslaai

AIZOACEAE   endemic

An annual, spreading herb up to 10cm
high. The fleshy, flat, inversely sword-
shaped leaves are up to 6cm long; the
water-storing epidermal cells are smaller
than those of *C. patersonjonesii*. The
yellow flowers are about 3cm in diameter.
As a seed-dispersal mechanism, the dry,
flat fruits have 5 valves that open when
wet and close again when the fruit dries.

Found in sandy soil in the Kamiesberg.
Flowers August to October.

ANNELISE LE ROUX

NICK HELME

There are about 100 species of *Antimima* in southern Africa of which 41 species occur in Namaqualand.

Antimima dualis

AIZOACEAE   endemic

A clump-forming perennial succulent up to 5cm high. The paired, fleshy, 3-angled, blue-green leaves, 2cm long, 5mm across and 5mm thick, are keeled and have convex to straight sides; the margin is hard and tough. The mauve flowers are 1.5cm in diameter; the filamentous staminodes and stamens are pale pink. As a seed-dispersal mechanism, the woody fruits have 5 valves that open when wet and close again when the fruit dries. Can easily be confused with *A. turneriana*.

Found in crevices on white quartz or limestone slopes on the Knersvlakte. **Flowers in June.**

ANNELISE LE ROUX

Antimima evoluta

AIZOACEAE

A rounded, clump-forming perennial succulent up to 5cm high. The small, paired, fleshy, blue-green leaves are covered with papillae, the margin with stiff, white hairs on the inside; the old leaves form a dry, white sheath around the new set of leaves in summer. The magenta-red flowers are up to 1.7cm in diameter. As a seed-dispersal mechanism, the woody fruits have 5 valves that open when wet and close again when the fruit dries.

Found in crevices on white quartz, marble or limestone slopes on the Knersvlakte, extending eastwards to Loeriesfontein.
Flowers in October.

Antimima fenestrata
AIZOACEAE   endemic

A compact shrub up to 4cm high, forming
tiny tree-like shapes. The leaves are tightly
packed on the short branches; the first
leaf pair consists of a long, smooth, white
sheath, embracing the subsequent leaves
at their base; the leaves end in a short
spine; the keel and angles are transparent;
the old leaves form a dry, white sheath
around the new set of leaves in summer.
The mauve flowers are 1.5cm in diameter.
As a seed-dispersal mechanism, the woody
fruits have 5 valves that open when wet
and close again when the fruit dries.
   Found in yellow, loamy soil or in
crevices on white quartz, marble or
limestone slopes on the Knersvlakte.
Flowers June to July.

ANNELISE LE ROUX

Antimima turneriana
AIZOACEAE   endemic

A clump-forming perennial succulent up
to 5cm high. The paired, fleshy, 3-angled,
blue-green leaves, 1.5cm long, 8mm across
and 8mm thick, are keeled and have
convex to straight sides; the margin is hard
and tough. The mauve flowers are 1.8cm
in diameter; the filamentous staminodes
and stamens are purple at the tip and
pale pink below. As a seed-dispersal
mechanism, the woody fruits have 5 valves
that open when wet and close again when
the fruit dries. Can easily be confused with
A. dualis.
   Found in crevices on white quartz,
marble or limestone slopes on the
Knersvlakte.
Flowers in June.

NICK HELME

ARGYRODERMA
There are 11 species of *Argyroderma* in southern Africa all of which are endemic to the Knersvlakte of Namaqualand.

Argyroderma delaetii
bababoudjies
AIZOACEAE   endemic

Single-headed succulent sunken into the ground. Leaves 2, very fleshy, greyish blue-green, almost elliptic but flattened above; the old leaves become grey when they wither and disintegrate during the first or second season. Flowers, up to 5cm in diameter, vary from white to yellow to pink-purple, the colours also mixed within the same population. As a seed-dispersal mechanism, the woody fruits have 14–24 valves that open when wet and close again when the fruit dries.

    Found among white quartz pebbles on the Knersvlakte.
Flowers April to June.

ANNELISE LE ROUX

Argyroderma fissum
vingervygie
AIZOACEAE   endemic

A clump-forming succulent up to 10cm high, with new branches appearing in the outer circle. The fleshy, cylindrical leaves, up to 6cm long and 1.8cm across, are grey-green, often tinted with yellow or purple at the ends. The flowers, up to 4.5cm in diameter, are pink-purple, often with a white centre; the filamentous staminodes are white. As a seed-dispersal mechanism, the woody fruits have 12 valves that open when wet and close again when the fruit dries.

    Found among white quartz pebbles on the Knersvlakte.
Flowers June to August.

ANNELISE LE ROUX & ADRIAN FORTUIN

Argyroderma framesii subsp. *framesii*

AIZOACEAE   endemic

A rounded, clump-forming perennial
succulent up to 5cm high. The 2 very
fleshy, grey-green leaves are at most 1cm
long. The deep mauve flowers are up
to 3cm in diameter. As a seed-dispersal
mechanism, the woody fruits have 10–12
valves that open when wet and close
again when the fruit dries.

Found among white quartz pebbles on
the Knersvlakte.
Flowers June to July.

ANNELISE LE ROUX

Argyroderma patens

AIZOACEAE   endemic

A clump-forming perennial succulent
up to 5cm high. The 2 very fleshy, grey-
green leaves are markedly keeled; the
gap between the 2 leaves is wider than
the visible remaining parts of the upper
leaf surface. The purple, mauve, white or
yellow flowers are up to 4cm in diameter.
As a seed-dispersal mechanism, the woody
fruits have 12 valves that open when wet
and close again when the fruit dries.

Found among white quartz pebbles on
the Knersvlakte.
Flowers May to June.

UTE SCHMIEDEL

BERNARD VAN LENTE

AMPHIBOLIA
There are five species of *Amphibolia* in southern Africa of which two occur in Namaqualand.

Amphibolia laevis
strandvygie
AIZOACEAE

A spreading to prostrate shrub up to 20cm high. The fleshy, club-shaped, green to grey-green leaves are about 1cm long and 5mm across; the old dry leaves remain on the branch for a long time. The flowers are pink to deep mauve. As a seed-dispersal mechanism, the woody, funnel-shaped fruits have 5 valves that open when wet and close again when the fruit dries.

Found spreading on the white coastal dunes in Namaqualand from just north of Hondeklip Bay, extending southwards to Simon's Town.
Flowers September to October.

JOANNE VAN SELM

ASTRIDIA
There are eight species of *Astridia* in southern Africa and in Namaqualand.

Astridia longifolia
AIZOACEAE   endemic

An erect shrub up to 35cm high. The fleshy, velvety, blue-green leaves, about 6cm long, are oblong and somewhat compressed laterally. Red flowers with a yellow to whitish centre and about 5cm in diameter are borne at the ends of branches; the stamens and staminodes are arranged in a cone. As a seed-dispersal mechanism, the woody fruits have 6 valves that open when wet and close again when the fruit dries.

Found on rocky hills and mountains in the Richtersveld.
Flowers August to September.

CARPOBROTUS
There are seven species of *Carpobrotus* in southern Africa of which two occur in Namaqualand.

Carpobrotus edulis
rankvy, veldvy, t'kôbôvy
AIZOACEAE

A creeping succulent with winged branches about 2m long. Flowering branches erect; with 2–4 leaf pairs. Leaves fleshy, straight or very slightly curved, tapering from about halfway to a point at the tip, triangular in cross-section and often tinged red. Flowers yellow turning pink with age, and up to 9cm in diameter. Fruits remain fleshy. The sweet-sour, fleshy fruit is eaten by humans and animals and is often used to make jam.

Found on sandy flats, often in dry riverbeds or along the coast, in Namaqualand, extending southwards along the coast to the Western and Eastern Cape.
Flowers August to October.

RIAAN DE VILLERS

Carpobrotus quadrifidus
elandsvy, suurvy
AIZOACEAE

A creeping perennial succulent with trailing branches up to 1m long, up to 2cm in diameter, 4–6-angled and winged. The very fleshy, bluish-green leaves are opposite, smooth and triangular in cross-section. The glistening pink flowers are up to 14cm in diameter. Fruits remain fleshy.

Found on the Coastal Plain, extending southwards to Langebaan.
Flowers August to September.

TESSA OLIVER

ANNELISE LE ROUX

CEPHALOPHYLLUM
There are 32 species of *Cephalophyllum* in southern Africa of which 18 occur in Namaqualand.

Cephalophyllum pillansii

AIZOACEAE   endemic

A prostrate succulent up to 10cm high. The fleshy, blue-green to bright green leaves are up to 6cm long. The yellow flowers have purple-red stamens. As a seed-dispersal mechanism, the woody fruits have 15–20 valves that open when wet and close again when the fruit dries; the ripe fruit stalks are long and, when dry, they lie on the ground in a ring around the plant.

Found in loamy soil on flats in Namaqualand.
Flowers June to September.

ZELDA WAHL

Cephalophyllum rigidum
(previously misidentified as *Cephalophyllum ebracteatum*)
rank-t'nouroe

AIZOACEAE   endemic

A creeping perennial succulent up to 5cm high. The fleshy, cylindrical, dull greyish-green leaves are up to 5cm long. The white, yellow, orange or pinkish flowers are about 5cm in diameter. As a seed-dispersal mechanism, the woody fruits, borne on persistent, decumbent stalks, have 8–15 valves that open when wet and close again when the fruit dries.

Found in loamy soil in Namaqualand, extending into Namibia.
Flowers June to August.

Cephalophyllum spissum

AIZOACEAE   endemic

A perennial succulent up to 10cm high.
The paired, fleshy leaves are up to 6cm
long and triangular in cross-section. The
flowers are reddish-pink, with a lighter
centre. As a seed-dispersal mechanism, the
woody fruits, borne on decumbent stalks,
have 11–15 valves that open when wet
and close again when the fruit dries.

Found in loamy soil among white
quartz pebbles on the Knersvlakte.
Flowers July to August.

ANNELISE LE ROUX

Cephalophyllum staminodiosum

AIZOACEAE   endemic

A tufted to creeping succulent up to 15cm
high. The fleshy, cylindrical, blue-green
leaves are blunt at the tip and up to 4.5cm
long. The white flowers have a column
of yellow filamentous staminodes with
stamens in the centre. As a seed-dispersal
mechanism, the woody fruits have 10
valves that open when wet and close
again when the fruit dries.

Found in loamy soil among white
quartz pebbles on the Knersvlakte.
Flowers June to August.

ANNELISE LE ROUX

CHEIRIDOPSIS
There are 29 species of *Cheiridopsis* in southern Africa of which 28 occur in Namaqualand.

Cheiridopsis denticulata
t'noutsiama

AIZOACEAE   endemic

A perennial succulent up to 15cm high. The fleshy, grey-green leaves are up to 11cm long and triangular in cross-section; the inside upper margin of some leaves are toothed. The flowers, up to 7cm in diameter, are white, cream-coloured to yellow, often rosy at the tips and on the outside; filamentous staminodes are absent. As a seed-dispersal mechanism, the woody fruits have 12–20 valves that open when wet and close again when the fruit dries.

Found on sandy flats in Namaqualand. **Flowers August to September.**

ZELDA WAHL & PIET SMITH

Cheiridopsis herrei
AIZOACEAE   endemic

A perennial succulent up to 5cm high. The fleshy, pale greyish-blue leaves, up to 3cm long, are triangular in cross-section and sharply keeled. The bright yellow flowers are up to 4cm in diameter. As a seed-dispersal mechanism, the woody fruits have 10 valves that open when wet and close again when the fruit dries.

Often found in spongy, saline soil among quartz pebbles or on rocky slopes in the Richtersveld.
Flowers August to September.

PIETER VAN WYK

Cheiridopsis imitans
AIZOACEAE   endemic

A small, compact perennial succulent up to 8cm high. The 2 pairs of opposite, fleshy, grey-green leaves are up to 5cm long and pointed to blunt at the tips; leaves of the previous season persist as a reddish-brown sheath during the dry season, protecting the younger inner pair. The yellow flowers are about 3cm in diameter. As a seed-dispersal mechanism, the woody fruits have 10 valves that open when wet and close again when the fruit dries.

Found on sandy plains with scattered rocks in the Namakwaland Klipkoppe. **Flowers August to September.**

ANNELISE LE ROUX

Cheiridopsis namaquensis
AIZOACEAE

A small, compact perennial succulent up to 8cm high. Two pairs of opposite, fleshy, grey-green leaves are up to 5cm long and triangular in cross-section with a velvety leaf surface covered with small papillae; leaves become dry, white and papery in summer, completely enclosing the new leaves and protecting them from drying out; this long, white, papery cover resembles a cigarette. Yellow flowers about 4.5cm in diameter. As a seed-dispersal mechanism, the woody fruits have 9 or 10 valves that open when wet and close again when the fruit dries.

Found in sandy soil in rocky areas in Namaqualand, extending eastwards to Calvinia and then southwards to Matjiesfontein in the Western Cape. **Flowers August to October.**

TESSA OLIVER & ANNELISE LE ROUX

Cheiridopsis peculiaris
eselore

AIZOACEAE   endemic

A small perennial succulent up to 5cm
high. There are 2 types of fleshy leaves: a
short pair and a longer pair, about 3 times
as broad as thick; the longer leaves are
almost shell-shaped, spread widely during
the rainy season (hence the vernacular
name 'eselore' meaning donkey's ears);
the short pair becomes dry, white and
papery in summer, enclosing the new
leaves completely and protecting them
from drying out. The yellow flowers are
up to 6cm in diameter. As a seed-dispersal
mechanism, the woody fruits have 14 or
15 valves that open when wet and close
again when the fruit dries.

Found in shale in the northern part of
the Namakwaland Klipkoppe.
Flowers August to September.

ZELDA WAHL

Cheiridopsis pillansii

AIZOACEAE   endemic

A succulent up to 8cm high, with many
leaf pairs. The fleshy, blue-green leaves
are broad and thick. The cream-coloured
to yellow flowers are sometimes tinged
purple. As a seed-dispersal mechanism, the
woody fruits, with a rounded base, have
10–14 valves that open when wet and
close again when the fruit dries.

Often found in spongy, saline soil
among quartz pebbles or on low slopes in
the Richtersveld.
Flowers August to September.

ANNELISE LE ROUX

Cheiridopsis robusta

AIZOACEAE   endemic

A perennial succulent up to 15cm high.
The opposite, fleshy, grey-green leaves
are up to 8cm long, triangular in cross-
section and have a single, short, stiff
point on the inner upper margin. The
flowers, up to 8cm in diameter, are yellow,
cream-coloured or white with a pink or
orange tinge; filamentous staminodes and
stamens are present. As a seed-dispersal
mechanism, the woody fruits have 10–13
valves that open when wet and close
again when the fruit dries.

   Found on rocky flats or slopes in the
Richtersveld and on the northern Coastal
Plain, extending into Namibia.
Flowers August to September.

PIETER VAN WYK

Cheiridopsis verrucosa
bokspoortjies
AIZOACEAE   endemic

A small, compact succulent up to 4cm
high, often covered by sand and therefore
appearing to be sunk into the ground.
The very fleshy, pale greyish-blue, spotted
leaves are rounded. The yellow flowers are
less than 3.5cm in diameter; filamentous
staminodes and stamens are present. As
a seed-dispersal mechanism, the woody
fruits, with a rounded base, have 10 valves
that open when wet and close again when
the fruit dries.

   Often found in sandy soil on shale or
dolomite, sometimes among scattered
quartz pebbles, on the very northern Coastal
Plain, extending into the south of Namibia.
Flowers August to September.

PIETER VAN WYK

CONOPHYTUM
There are 87 species of *Conophytum* in southern Africa of which 79 occur in Namaqualand.

Conophytum bilobum
AIZOACEAE   endemic

A clump-forming perennial succulent up to 5cm high. The fleshy, partially fused leaf pair forms a V-shape and is yellowish-green to whitish-green, with a red line on the keel and into the fork. The yellow, rarely white or pink flowers usually have a long tube. As a seed-dispersal mechanism, the dry fruits usually have 4 valves that open when wet and close again when the fruit dries.

Found among granite or quartz rocks in the Richtersveld and Namakwaland Klipkoppe.
Flowers February to June.

NORBERT JÜRGENS

Conophytum calculus
AIZOACEAE

A clump-forming perennial succulent up to 3cm high. The firm, fleshy, totally fused leaf pair is typically round, chalky blue-green, with a fissure in the middle; the leaves of the past winter become white and papery in summer, enclosing the leaves of the next season and protecting them from drying out. The flowers, golden to orange, often tipped with red, are strongly carnation-scented and open at night. As a seed-dispersal mechanism, the dry fruits have 5–7, usually 6, valves that open when wet and close again when the fruit dries.

Found among white quartz pebbles, often on slight slopes, in the Namakwaland Klipkoppe and on the Knersvlakte, extending eastwards into Bushmanland.
Flowers April to June.

KOBUS KRITZINGER

Conophytum concavum
AIZOACEAE   endemic

A clump-forming perennial succulent up
to 5mm high and sunk into the ground.
The fleshy leaves are depressed (concave)
towards the centre and have a velvety
texture; the leaves dry out and become
white and papery in summer, enclosing
the new leaves completely and protecting
them from drying out. The pale yellow to
white flowers are honey-scented and open
in the late afternoon. As a seed-dispersal
mechanism, the dry fruits usually have
5 valves that open when wet and close
again when the fruit dries.
    Found among white quartz pebbles on
the eastern Coastal Plain.
Flowers April to May.

CARLO VAN TONDER

Conophytum khamiesbergense
AIZOACEAE   endemic

A clump-forming perennial succulent up
to 3cm high. The fleshy, wedge-shaped,
greyish-green to brownish leaves are
covered in whitish protuberances giving
the leaf a warty appearance; the leaves
give off a raspberry scent. The flowers
are whitish-pink to mauve and sweetly
scented of raspberries and vanilla;
filamentous staminodes and stamens are
present. As a seed-dispersal mechanism,
the dry fruits usually have 4 valves that
open when wet and close again when the
fruit dries.
    Found in rock crevices on mountain
slopes in the Kamiesberg of the Namakwaland Klipkoppe.
Flowers August to October.

ANNELISE LE ROUX

Conophytum pellucidum subsp. *pellucidum*
bruintandjies
AIZOACEAE   endemic

A clump-forming perennial succulent up to 5mm high and sunk into the ground. The fleshy, brown leaves are flattened and have darker brown, irregular grooves and sunken spots; the leaves dry out and become white and papery in summer, enclosing the new leaves completely and protecting them from drying out. The white flowers have a long tube. As a seed-dispersal mechanism, the dry fruits usually have 5 valves that open when wet and close again when the fruit dries.

Found in shallow, sandy depressions on large, flat, granite boulders in the Namakwaland Klipkoppe.
Flowers February to April.

Conophytum subfenestratum
AIZOACEAE   endemic

A solitary or sometimes clump-forming perennial succulent up to 3cm high. The fleshy, spotted leaf pair is fused and up to 2cm high above ground; the leaves dry out and become white and papery in summer, enclosing the new leaves completely and protecting them from drying out. The pale pink to mauve flowers are slightly scented. As a seed-dispersal mechanism, the dry fruits have 4–7, usually 4, valves that open when wet and close again when the fruit dries.

Found in loamy soil among white quartz pebbles on the Knersvlakte.
Flowers March to April.

DICROCAULON
There are seven species of *Dicrocaulon* in southern Africa and in Namaqualand.

Dicrocaulon ramulosum
AIZOACEAE

An erect shrub up to 20cm high. Two pairs of fleshy leaves are borne on a short branch per season: the first leaf pair is fused into a round body and is shed before flowering starts; the cylindrical leaves of the second leaf pair are fused only at the bottom and are up to 5.5cm long, with a blunt end. White flowers are up to 3.5cm in diameter. As a seed-dispersal mechanism, the dry fruits have 4–6, usually 5, valves that open when wet and close again when the fruit dries.

Found in loamy soil among white quartz pebbles on the Knersvlakte or in sandy soil in flat depressions of rocks on the Coastal Plain.
Flowers July to August.

UTE SCHMIEDEL

DIPLOSOMA
There are two species of *Diplosoma* in southern Africa of which one occurs in Namaqualand.

Diplosoma luckhoffii
AIZOACEAE   endemic

A deciduous perennial succulent up to 2.5cm high. Two pairs of very fleshy leaves: the first pair almost round, emerging at the beginning of the growing season; the oblong leaves of the second pair up to 2cm long and 1cm in diameter, with the upper surface slightly concave; water-storing epidermal cells large and conspicuous. Flowers mauve with a white centre; all or at least the outer filamentous staminodes dark purple, nearly black. As a seed-dispersal mechanism, the fruits usually have 6 valves that open when wet and close again when the fruit dries.

Found in loamy soil among white quartz pebbles on the Knersvlakte.
Flowers June to August.

PATRICK LANE

AIZOACEAE **317**

UTE SCHMIEDEL

DROSANTHEMUM
There are about 110 species of *Drosanthemum* in southern Africa of which 17 occur in Namaqualand.

Drosanthemum diversifolium
AIZOACEAE

A spreading shrub up to 20cm high. Branches have small spines; papillae on young branches become white dots on a reddish-brown bark when older and dry. Leaves fleshy, boat-shaped, up to 1cm long, 4mm across and 5mm thick; water-storing epidermal cells are cobblestone-shaped, those on the leaf edge finger-like. Flowers pink and borne on flower stalks with erect hairs; filamentous staminodes absent. Woody fruits have 6 valves that open when wet and close again when the fruit dries.

Found in red loamy soil, sometimes among white quartz pebbles, in Namaqualand, extending southwards to Clanwilliam. **Flowers July to September.**

ZELDA WAHL

Drosanthemum hispidum
fyn t'nouroebos
AIZOACEAE

A spreading shrub up to 20cm high. Red branches are covered with white bristles. The fleshy, cylindrical leaves are blunt at the tip and up to 2cm long; the water-storing epidermal cells are densely arranged, soft and glistening. The mauve flowers are up to 3cm in diameter; filamentous staminodes are absent. As a seed-dispersal mechanism, the woody fruits, with a funnel-shaped base and flat top, have 5 valves that open when wet and close again when the fruit dries.

Found in sandy soil, mostly on flats, in Namaqualand, extending into other dry areas of the Northern, Western and Eastern Cape. **Flowers August to September.**

Drosanthemum ramosissimum
AIZOACEAE   endemic

A spreading shrub up to 20cm high. Pale
yellow-brown branches are covered with
rough, round papillae. The fleshy, cylindrical
leaves are blunt at the tip and up to 1cm
long; the large, soft, glossy, water-storing
epidermal cells are densely arranged in a
thick, bluish layer. The pale to dark mauve
flowers are up to 2cm in diameter. As a
seed-dispersal mechanism, the woody fruits
have 5 valves that open when wet and close
again when the fruit dries.

Found in loamy soil among white
quartz pebbles on the Knersvlakte.
Flowers August to September.

UTE SCHMIEDEL

Drosanthemum salicola
AIZOACEAE

A dense, spreading shrub up to 25cm
high. The fleshy, round leaves are up
to 7mm in diameter; the round, glossy,
water-storing epidermal cells are densely
arranged. The pale pink flowers are
up to 2cm in diameter; the numerous
filamentous staminodes are pale pink;
the stamens have pale pink filaments
and white anthers. As a seed-dispersal
mechanism, the woody fruits have 5
valves that open when wet and close
again when the fruit dries.

Found in sandy soil among rocks
very close to the sea on the Coastal
Plain from Alexander Bay extending
southwards to Velddrift.
Flowers October to December.

HEATHER BURGER

Drosanthemum schoenlandianum
AIZOACEAE   endemic

A spreading to prostrate shrub up to 15cm high, with stiff long hairs on the branches. The fleshy, cylindrical leaves are blunt at the tip and up to 8mm long; the round, water-storing epidermal cells are glistening. White or cream- to lemon-coloured flowers are borne on stalks with long, stiff hairs pointing downwards and shorter hairs among them. As a seed-dispersal mechanism, the woody, bell-shaped fruits have 5 valves that open when wet and close again when the fruit dries.

Found in red loamy soil on the Knersvlakte.
Flowers July to September.

ANNELISE LE ROUX

EBERLANZIA
There are eight species of *Eberlanzia* in southern Africa of which seven occur in Namaqualand.

Eberlanzia dichotoma
AIZOACEAE   endemic

An erect shrub up to 25cm high. The fleshy, blue-green leaves are short and club-shaped. Numerous small, red-purple flowers are borne in clusters on purple stalks at the ends of branches; the filamentous staminodes are purple with white bases. The woody fruits are up to 6mm in diameter and, as a seed-dispersal mechanism, have 5 valves that open when wet and close again when the fruit dries.

Found in loamy soil on 'heuweltjies' on the Coastal Plain.
Flowers August to October.

ANNELISE LE ROUX

There is one species with two subspecies of
Fenestraria in southern Africa of which one
subspecies occurs in Namaqualand.

Fenestraria rhopalophylla subsp. *aurantiaca*

window plant, vensterplantjie

AIZOACEAE

A dwarf perennial succulent up to 5mm
above ground in winter, completely underground in summer. Fleshy leaves covered by
a wax layer; upper surface just above the
soil has a translucent 'window' that allows
sunlight through to the underground, green
photosynthesising part of leaf. Flowers
yellow to pale orange, up to 5cm across.
Fruits have up to 16, usually 10, valves that
open when wet and close when fruit dries.

    Found in windblown sand in the fog
zone on the coast from north of the
Orange River to south of Kleinzee.
Flowers May to July.

PIETER VAN WYK

HALLIANTHUS
There are two species of *Hallianthus* in southern
Africa and both occur in Namaqualand.

Hallianthus planus

AIZOACEAE   endemic

A prostrate shrub with a central tuft and
trailing side branches. Reddish-brown
young branches do not root at the nodes.
The fleshy, dark green leaves are triangular
to sickle-shaped. The flowers are white or
mauve; the filamentous staminodes and
stamens form a central column. Fruits are
borne on long, prostrate stalks and, as a
seed-dispersal mechanism, have 10 valves
that open when wet and close again when
the fruit dries.

    Found in sandy soil on flats and rocky
slopes in Namaqualand.
Flowers June to August.

ANNELISE LE ROUX

CAROLIN MAYER

Ihlenfeldtia excavata
AIZOACEAE

A compact succulent up to 5cm high. The fleshy, triangular, blue-green leaves have a short, stiff point at the tip and often have small teeth. The flowers, up to 7cm in diameter, are various shades of yellow to white with reddish tints, sometimes orange or purplish. As a seed-dispersal mechanism, the woody fruits, with a funnel-shaped base, have 10–15 valves that open when wet and close again when the fruit dries.

Found in sandy soil, often among white quartz pebbles, in the eastern part of the Namakwaland Klipkoppe, extending slightly into Bushmanland. **Flowers August to September.**

ANNELISE LE ROUX

JORDAANIELLA
There are seven species of *Jordaaniella* in southern Africa of which four occur in Namaqualand.

Jordaaniella clavifolia
AIZOACEAE   endemic

A creeping shrub up to 5cm high. The internodes of the branches are up to 2cm long and 5mm in diameter. The fleshy, cylindrical leaves are up to 5cm long and 5mm across, the broadest part usually more than halfway up and hence the leaves are club-shaped; the leaves droop on the ground during the dormant period The flowers are yellow and up to 7.5cm in diameter. As a seed-dispersal mechanism, the woody fruits have 14 or 15 valves that open when wet and close again when the fruit dries.

Found in sandy soil on flats on the Coastal Plain from Kleinzee to Hondeklip Bay. **Flowers June to July.**

322 AIZOACEAE

Jordaaniella cuprea
AIZOACEAE

A creeping shrub up to 15cm high.
The internodes of the branches are
less than 1.5cm long and more than
5mm in diameter. The fleshy, cylindrical
leaves are up to 10cm long and 1.4cm
across; the leaves are erect during the
dormant period. The flowers, up to 10cm
in diameter, are pale to dark yellow,
with orange to pink tips, often salmon-
coloured. As a seed-dispersal mechanism,
the woody fruits have 14–20 valves that
open when wet and close again when the
fruit dries.

    Found in sandy soil on flats on the
northern Coastal Plain, extending into
Namibia.
Flowers June to August.

PIETER VAN WYK

Jordaaniella spongiosa
volstruisvygie

AIZOACEAE   endemic

A shrub up to 30cm high. Branches are up
to 1cm in diameter. The fleshy, cylindrical,
bluish-green leaves are up to 10cm long.
The flowers, up to 10cm in diameter, are
apricot to cerise with a yellow to orange
centre. As a seed-dispersal mechanism,
the fruits, with a short funnel-shaped
base and flat top, have 18–28 valves that
open when wet and close again when the
fruit dries.

    Found in sandy soil on the Coastal Plain.
Flowers August to September.

ZELDA WAHL

LAMPRANTHUS
There are about 90 species of *Lampranthus* in southern Africa of which 12 occur in Namaqualand.

Lampranthus otzenianus
AIZOACEAE

A cushion-shaped shrub up to 40cm high. The fleshy, cylindrical, slightly curved, yellow-green leaves are up to 2.5cm long. The bright mauve flowers are up to 3cm in diameter; filamentous staminodes are absent. The greyish-white fruits are bell-shaped and, as a seed-dispersal mechanism, have 5 valves that open when wet and close again when the fruit dries.

Found in loamy soil on flats, mostly on 'heuweltjies', extending into Bushmanland, on to the Tankwa Karoo and into Namibia. **Flowers August to September.**

ANNELISE LE ROUX

Lampranthus stipulaceus
(previously *misidentified as Lampranthus suavissimus*)
vygiebos, langbeen-t'nouroebos
AIZOACEAE

An erect shrub up to 75cm high. The fleshy, cylindrical, bluish-green leaves are up to 3.5cm long. The pink, sometimes white flowers are up to 5.5cm in diameter; filamentous staminodes are absent. As a seed-dispersal mechanism, the woody fruits have 5 valves that open when wet and close again when the fruit dries.

Found in deep, sandy soil on the Coastal Plain, extending southwards to Cape Town and northwards into Namibia. **Flowers August to September.**

ZELDA WAHL

Lampranthus watermeyeri
AIZOACEAE

A spreading to erect shrub up to 30cm high. The fleshy, almost cylindrical leaves are up to 2.5cm long. The white flowers are up to 6cm in diameter; filamentous staminodes are absent. As a seed-dispersal mechanism, the woody fruits have 5 valves that open when wet and close again when the fruit dries.

Found in sandy soil on rocky slopes, sometimes among white quartz pebbles, in Namaqualand, extending southwards to Lambert's Bay and southeastwards to the Tankwa Karoo.
Flowers July to November.

ANNELISE LE ROUX

Leipoldtia schultzei
langbeen-t'nouroebos
AIZOACEAE

A spreading shrub up to 40cm high. The fleshy leaves are triangular in cross-section, up to 2.5cm long and have raised dots on the leaf surface. Bright mauve flowers up to 3cm in diameter are borne in groups of 3–5 towards the ends of branches; filamentous staminodes are absent. As a seed-dispersal mechanism, the woody fruits, with a funnel- to bell-shaped base and flat top, have 10–12 valves that open when wet and close again when the fruit dries.

Found in loamy soil on flats and rocky slopes in Namaqualand, extending into the Western and Eastern Cape.
Flowers August to September.

ZELDA WAHL

AIZOACEAE **325**

MEYEROPHYTUM
There are two species of *Meyerophytum* in southern
Africa and both occur in Namaqualand.

Meyerophytum globosum
AIZOACEAE   endemic

A compact succulent up to 10cm high and
20cm in diameter. There are 2 types of
fleshy leaves: the first leaf pair is round,
up to 1.7cm long and 1.4cm across,
while the second pair is cylindrical and
up to 7cm long with the 2 leaves basally
fused for up to 1.5cm; old leaves wither
completely into a white papery covering
in summer, protecting the round young
buds (see insert). Flowers are white with
a yellow centre. Woody fruits have 5,
sometimes 6, valves that open when wet
and close again when the fruit dries.
    Found in loamy soil among quartz
pebbles in the eastern part of the
Coastal Plain.
Flowers August to September.

NICK HELME & IVAN VAN NIEKERK

Meyerophytum meyeri
AIZOACEAE

A cushion-shaped shrub up to 10cm high
and 30cm in diameter. There are 2 types
of fleshy leaves: the first leaf pair, up
to 1.2cm long and 1cm across, is almost
round; the second pair, up to 3.5cm long,
is basally fused; the old leaves wither
completely into a white papery covering in
summer, protecting the round young buds
(see insert). The flowers are deep pink-
mauve, sometimes with a white centre, or
white turning pink with age; the filaments
are purple, yellow or white. As a seed-
dispersal mechanism, the woody fruits
have 4–7, usually 5, valves that open when
wet and close again when the fruit dries.
    Found in loamy soil, often among
white quartz pebbles, on the northern
Coastal Plain.
Flowers July to August.

PIETER VAN WYK

There are 17 species of *Malephora* in southern Africa of which one occurs in Namaqualand.

Malephora purpureo-crocea

AIZOACEAE   endemic

A spreading shrub up to 30cm high. Branches are grey-brown and spongy. The fleshy, blue-green leaves are cylindrical to slightly triangular with a blunt end. The brilliant red flowers are about 3cm in diameter. As a seed-dispersal mechanism, the dry, soft, light brown fruits have 9 or 10 valves that open when wet and close again when the fruit dries.

Found in loamy soil in the Namakwaland Klipkoppe and on the Knersvlakte. **Flowers August to September.**

ANNELISE LE ROUX

There are six species of *Mitrophyllum* in southern Africa of which all are endemic to Namaqualand.

Mitrophyllum dissitum

AIZOACEAE   endemic

An erect shrub up to 50cm high. There are 2 types of fleshy, blue-green leaves: the first leaf pair, up to 5cm long and 1.5cm across, is tongue-shaped to triangular; the second leaf pair, up to 4cm long and 1.5cm across, is fused for four-fifths of its length. The flowers are yellow, sometimes tinted pink. As a seed-dispersal mechanism, the woody fruits have 5 valves that open when wet and close again when the fruit dries.

Found in loamy soil, usually in crevices on southern and southeastern slopes in the Richtersveld. **Flowers May to June.**

PIETER VAN WYK

AIZOACEAE **327**

MONILARIA
There are five species of *Monilaria* in southern Africa of which all are endemic to Namaqualand.

Monilaria chrysoleuca
AIZOACEAE   endemic

A spreading shrub up to 15cm high. Branches are covered with sheaths that are bell-shaped and the rim often irregularly lacerated at the nodes. There are 2 types of fleshy leaves: the first leaf pair is fused to form a nearly spherical body up to 1cm in diameter; the leaves of the second pair are up to 10cm long and 5mm across. The flowers are white, yellow, orange, salmon, red or purple; the filaments are sometimes a different colour, even if the petals are pure white. As a seed-dispersal mechanism, the woody fruits have 5 valves that open when wet and close again when the fruit dries.

Found in loamy soil among white quartz pebbles on the Knersvlakte. **Flowers July to August.**

ANNELISE LE ROUX

Monilaria moniliformis
AIZOACEAE   endemic

A spreading shrub up to 15cm high. Branches have flattish and rounded internodes and the nodes are constricted. There are 2 types of fleshy leaves: the first leaf pair is fused, flattish and round; the leaves of the second pair, up to 10cm long, are almost triangular in cross-section. The flowers are white, sometimes tinged yellow. As a seed-dispersal mechanism, the woody fruits have 5 valves that open when wet and close again when the fruit dries.

Found in loamy soil among white quartz pebbles on the Knersvlakte. **Flowers July to August.**

ANNELISE LE ROUX

Monilaria obconica

AIZOACEAE   endemic

A spreading shrub up to 15cm high.
Branches have flattish and rounded
internodes and the nodes are constricted.
There are 2 types of fleshy leaves: the
first leaf pair is fused and round; the
leaves of the second pair, up to 10cm
long, are cylindrical with a rounded,
often reddish tip. The white flowers, up
to 4cm in diameter, have bright yellow
stamens in the centre. As a seed-dispersal
mechanism, the woody fruits have 5
valves that open when wet and close
again when the fruit dries.

   Found in rocky, reddish soil in the
Namakwaland Klipkoppe.
Flowers July to August.

HEATHER BURGER

NELIA
There are two species of *Nelia* in southern Africa and
both are endemic to Namaqualand.

Nelia pillansii

AIZOACEAE   endemic

A tufted shrub up to 10cm high. The fleshy,
grey-green leaves are slightly triangular in
cross-section and a have rounded tip. The
white to yellowish flowers are up to 1.5cm
in diameter; the petals are stiff and merge
into the filamentous staminodes; unlike in
most other vygies or mesems, the flowers
are permanently open and do not close
at night; the dead, dry, stiff, black petals
surround the fruit for a long time. As a
seed-dispersal mechanism, the woody fruits
have 5 valves that open when wet and
close again when the fruit dries.

   Found in loamy soil covered with
quartz pebbles in the eastern part of the
Coastal Plain.
Flowers August to September.

ANNELISE LE ROUX

SOPHIA EITZOLD

Oophytum nanum
AIZOACEAE   endemic

A compact succulent up to 3cm high. There are only 2 fleshy, slightly keeled leaf pairs per season and both are visible; the green leaves quickly turn red; the old leaves wither completely into a white papery covering in summer, protecting the round young buds; the white leaves give the appearance of a cluster of small eggs among the white quartz pebbles. The flowers are deep mauve with a white centre; they open late in the afternoon for a short period only. The dry fruits have 6 valves that open when wet and close again when the fruit dries.

Found in loamy soil among white quartz pebbles on the Knersvlakte. **Flowers August to September.**

ANNELISE LE ROUX

Oophytum oviforme
krapogies
AIZOACEAE   endemic

A compact succulent up to 5cm high. There are only 2 fleshy leaves per season; the second leaf pair of the season is included in the first united ovoid pair; the old leaves wither completely into a white papery covering in summer, protecting the round young buds; the white leaves give the appearance of a cluster of eggs among the white quartz pebbles. The flowers are mauve with a white centre. The dry fruits are longer than broad and, as a seed-dispersal mechanism, have 6 valves that open when wet and close again when the fruit is dry.

Found in loamy soil among white quartz pebbles on the Knersvlakte. **Flowers August to September.**

ODONTOPHORUS
There are four species of *Odontophorus* in southern
Africa and all are endemic to Namaqualand.

Odontophorus marlothii

AIZOACEAE   endemic

A prostrate shrub up to 8cm high. Long,
dark reddish-brown branches develop
longitudinal folds with age. The fleshy
leaves, up to 3.5cm long and 1cm across,
are triangular in cross-section; the inner
margin has conspicuous long teeth. The
yellow flowers are up to 4cm in diameter.
As a seed-dispersal mechanism, the dry
fruits have 10 valves that open when wet
and close again when the fruit is dry.
     Found in sandy soil on rocky slopes in
the Namakwaland Klipkoppe.
Flowers July to August.

HEATHER BURGER

PSAMMOPHORA
There are four species of *Psammophora* in southern
Africa of which three occur in Namaqualand.

Psammophora modesta

gomvy

AIZOACEAE

A small shrub up to 15cm high. Branches
spread horizontally at first but then grow
upwards. Leaves fleshy, up to 4cm long,
1cm across and 8mm thick, longer than the
internodes and triangular in cross-section,
rarely roundish; leaf surface has a rough
appearance, with sand adhering to it.
Flowers, up to 2.5cm in diameter, and pink
to purple, sometimes with a white centre;
stamens numerous. Fruits have a spongy
base and have 4–9 valves that open when
wet and close again when the fruit dries.
     Found on the Coastal Plain near
the Orange River, extending into
southern Namibia.
Flowers June to September.

AIZOACEAE **331**

PIETER VAN WYK

RUSCHIA
There are about 220 species of *Ruschia* in southern Africa of which 55 occur in Namaqualand.

Ruschia aggregata
vygiebos, t'nouroebos
AIZOACEAE   endemic

An erect shrub up to 75cm high. The fleshy, cylindrical, blue-green leaves are up to 2.5cm long. Magenta-pink flowers are borne in clusters at the ends of branches; filamentous staminodes are absent; anthers and pollen are yellow-brown. The woody fruits have a funnel-shaped base, a raised top, low rims and, as a seed-dispersal mechanism, have 5 valves that open when wet and close again when the fruit dries.

Found on sandy flats or rocky slopes in the Namakwaland Klipkoppe and on the Knersvlakte.
Flowers August to October.

Ruschia muelleri
vygiebos, t'nouroebos
AIZOACEAE

An erect shrub up to 50cm high. Branches are shiny, spongy and white to brownish-yellow. The fleshy, blue-green leaves, up to 4cm long, are cylindrical with a blunt tip. Magenta-pink flowers are borne in clusters at the ends of branches; the filamentous staminodes have dark tips. The woody fruits have a markedly bell-shaped base, a round top reaching over the outer edge and, as a seed-dispersal mechanism, have 5 valves that open when wet and close again when the fruit dries.

Found on sandy, stony flats in Namaqualand, extending into Namibia.
Flowers July to August.

Ruschia robusta
swart t'nouroebos, swartstamvyebos
AIZOACEAE

An erect shrub up to 1m high, with dark brown to black branches. The fleshy, blue-green leaves, about up to 5mm in diameter and 5mm long, are shortly oblong to round; the leaf surface has translucent dots. Pink-purple flowers about 1.5cm in diameter are borne singly on very short branches all along the main branches. As a seed-dispersal mechanism, the woody fruits have 5 valves that open when wet and close again when the fruit dries.

Found on sandy flats on the eastern side of the Namakwaland Klipkoppe, extending into Bushmanland.
Flowers August to September.

ZELDA WAHL

Ruschia viridifolia
rankvygie, rank t'nouroe
AIZOACEAE

A prostrate shrub with branches up to 35cm long. The fleshy, bright green leaves, up to 4cm long, are cylindrical with a sharp tip. Magenta-pink flowers are borne in clusters of 3–5; filamentous staminodes are absent; the filaments are purple. The woody fruits have a funnel- to bell-shaped base and, as a seed-dispersal mechanism, have 5 valves that open when wet and close again when the fruit dries.

Found in rock crevices on flat rock sheets in the Richtersveld and Namakwaland Klipkoppe.
Flowers September to October.

UTE SCHMIEDEL

ZELDA WAHL

STOEBERIA
There are six species of *Stoeberia* in southern Africa
of which five occur in Namaqualand.

Stoeberia beetzii

vaal t'kooi

AIZOACEAE   endemic

An erect to spreading shrub up to about
60cm high. The fleshy, grey-green leaves,
up to 2cm long, are club-shaped; the
leaf surface has translucent spots. White
flowers about 1cm in diameter are borne
in clusters extending above the plant. The
woody fruits have 5 valves that remain
open after they have been wet.

Found in loamy soil along the northern
Coastal Plain and in Namibia.

Flowers August to September.

ANNELISE LE ROUX

Stoeberia frutescens

donkievy

AIZOACEAE

A spreading to erect shrub up to 75cm
high. The fleshy, dark bluish-green leaves,
up to 5cm long, are linear to slightly
triangular in cross-section. White flowers,
up to 1.2cm in diameter and flushed pink,
are borne in loose clusters at the end of a
branch. The woody fruits have 5 valves that
remain open after they have been wet.

Found in sandy soil, often on
'heuweltjies', in Namaqualand and also in
southern Namibia.

Flowers August to September.

Stoeberia utilis
t'arra-t'kooi, rooi-t'kooi
AIZOACEAE

An erect shrub up to 1.5m high. The fleshy, dark bluish-green leaves, up to 1.5cm long, are triangular in cross-section. White flowers about 1cm in diameter are borne singly on very short branches along the main branches. The woody fruits are hard and have 5 valves that remain open after they have been wet.

Found in loose sand along a 10km coastal belt on the Coastal Plain, extending southwards to Velddrift. **Flowers August to October.**

WOOLEYA
There is one species of *Wooleya* in southern Africa and it occurs in Namaqualand.

Wooleya farinosa
vaalvygie
AIZOACEAE

A moderately branched, sprawling shrub up to 30cm high. Young branches are ash-grey to purplish, becoming grey-brown. The fleshy, ash-grey to purplish leaves, up to 3cm long and 1cm across, are obscurely 3-angled to cylindrical, with the upper surface slightly flattened and the lower surface convex. The white flowers have numerous stamens, with few staminodes gathered into a cone around the stamens. The dry, woody fruits have a convex top and 11 or 12 valves that open when wet and close again when the fruit dries.

Found in sandy soil on the Coastal Plain between Port Nolloth and Hondeklip Bay. **Flowers March to August.**

AIZOACEAE **335**

ANNELISE LE ROUX

Kissenia capensis
LOASACEAE

A shrub up to 1m high, with white stems. The egg-shaped to oblong leaves are divided into 3–5 lobes; the margin is irregularly toothed. The flowers are white; the calyx tube is densely hairy and the 5 oblong lobes, about 4cm long, are pale green, lengthening and turning yellow to light brown after the flower has died.

Found in sandy soil among boulders on riverbanks and in dry watercourses in the Richtersveld, extending eastwards into Gordonia and northwards into Namibia. **Flowers February to October.**

ANNELISE LE ROUX

DIOSPYROS
There are about 20 species of *Diospyros* in southern Africa of which four occur in Namaqualand.

Diospyros austro-africana
kraaibos
EBENACEAE

An erect to spreading shrub up to 2.5m tall, with ash-grey to blackish bark. The stiff and leathery, broadly linear leaves, up to 1.5cm long, are densely covered with hairs, giving the plant a grey appearance. The cream-coloured to pink flowers droop and are densely hairy. The round fleshy fruits turn red to black when mature.

Found on rocky flats and slopes on the Coastal Plain and in the Namakwaland Klipkoppe, the Bokkeveld, Hantam and Roggeveld Mountains and extending from the Western Cape to Mpumalanga. **Flowers January to August.**

Diospyros ramulosa
t'koenoebe
EBENACEAE

An erect, much-branched shrub up to
1.8m high, with dark grey stems. The
small leaves are elliptic, entire, crowded at
the ends of branches and silky-haired on
the underside. The flowers are small and
greenish-white to cream-coloured. The
orange to red fruits are round to slightly
elliptic, 1–2cm in diameter and hairy. A
very palatable plant to stock; the large red
fruits are eaten by humans and animals.

Found on rocky slopes and sandy
flats throughout the Richtersveld,
Namakwaland Klipkoppe and Coastal
Plain, extending southwards to
Clanwilliam and northwards into Namibia.
Flowers April to September.

ANNELISE LE ROUX

EUCLEA
There are 17 species of *Euclea* in southern Africa of
which five occur in Namaqualand.

Euclea lancea
Namaqua guarri, Namakwa-gwarrie
EBENACEAE

A much-branched, evergreen shrub or small
tree up to 3m tall, with greenish-brown
bark. The sharply pointed, linear-elliptic,
thickly leathery leaves are blue-green above
and paler below, with a thickened margin.
The drooping flowers are cream-coloured
and have 15–22 stamens.

Found on rocky slopes near
watercourses in the Namakwaland
Klipkoppe and from the Bokkeveld
Mountains to the Cederberg.
Flowers October to April.

ANNELISE LE ROUX

Euclea pseudebenus
black ebony, swartebbehout
EBENACEAE

A small, evergreen tree up to 4m tall, with rough dark bark and slender drooping branches. The leathery, bluish-green leaves are very long, slender, slightly curved and strongly aromatic. Small, greenish-yellow flowers with 12–20 stamens are borne between the leaves. The round fruits become black when mature.

Found along dry riverbeds and watercourses in the Richtersveld, northern Bushmanland and in Namibia. **Flowers March to May.**

Euclea racemosa
dune guarri, duingwarrie
EBENACEAE

A large, much-branched, evergreen shrub up to 3m tall, with a smooth, grey bark. The leathery, inversely egg-shaped, yellow-green leaves have a thickened margin. The fragrant cream-coloured flowers have 10–20 stamens. Fruits are round and hairy.

Found in sand on the Coastal Plain and along the coasts of the Western Cape, the Eastern Cape and Namibia. **Flowers December to June.**

Euclea tomentosa
klipkers
EBENACEAE

A much-branched, evergreen shrub up to 1.5m high, with blackish-grey bark. The inversely egg-shaped, leathery leaves are covered with hairs when young, but lose these hairs with age. The fragrant white flowers have 16–21 stamens. The fleshy, round fruits turn red to black when mature.

Found on rocky slopes and in ravines in the Richtersveld, Namakwaland Klipkoppe and in the Western Cape.
Flowers March to October.

ANNELISE LE ROUX

Euclea undulata
guarri, gwarrie
EBENACEAE

A much-branched, evergreen shrub or small tree up to 3.5m tall, with grey, scaly bark. The firm, inversely sword-shaped leaves are often dark green above and paler below; the margin is undulate. The fragrant cream-coloured flowers have 12–20 stamens. The fleshy, round fruits turn red to purple when mature. The leaves are browsed by cattle.

Found on rocky slopes and in dry riverbeds in the Namakwaland Klipkoppe, on the Knersvlakte and also widespread elsewhere in southern Africa, extending to tropical Africa.
Flowers December to April.

ANNELISE LE ROUX

ZELDA WAHL

ERICA
There are over 750 species of *Erica* in southern Africa
of which eight occur in Namaqualand.

Erica plukenetii
hangertjieheide
ERICACEAE

An erect shrub up to 1m high. The leaves
are up to 8mm long, linear, spreading
or recurved. Bright red, tubular flowers
about 2.5–3cm long are borne in clusters
on the branches; the anthers are exserted
from the flower tube.

Found on rocky mountain slopes in
the Spektakelberg and Kamiesberg in
Namaqualand and extending from the
Bokkeveld Mountains to the Langeberg.
Flowers August to October.

IVAN VAN NIEKERK

Erica verecunda
heide
ERICACEAE

A shrub up to 80cm high. The leaves,
borne on red branches, are arranged in
distinct whorls of 4 and have sharp points.
White to pinkish, bell-shaped flowers up
to 4mm long are borne on 5mm-long, red
stalks; the anthers are not exserted from
the corolla.

Found on dry rocky mountain
slopes, usually along seepages, in the
Namakwaland Klipkoppe and Bokkeveld
Mountains, extending to near Touwsrivier.
Flowers December to May.

CHIRONIA
There are 16 species of *Chironia* in southern Africa of
which two occur in Namaqualand.

Chironia baccifera
toothache berry, maagbitters,
tandpynbossie
GENTIANACEAE

A rounded shrub up to 1m high, with
angled stems. The small, narrow leaves
spread out in different directions; the
margin is rolled back. The flowers are
shiny pink. The fleshy, round fruits become
bright red when ripe.

Found on sand dunes and on rocky
slopes in the Namakwaland Klipkoppe, on
the Coastal Plain and also in the Western
Cape, extending further along the coastal
belt into KwaZulu-Natal.
Flowers July to December.

ANNELISE LE ROUX

Chironia linoides
GENTIANACEAE

A sparse, erect or spreading shrub up to
90cm high. The narrow leaves spread out
in different directions; the margin of the
basal half, but not the upper half, is rolled
back, giving the leaves a spoon-shaped
appearance. Shiny pink flowers are borne
in sparse terminal clusters.

Found in sandy or marshy areas, on
mountains, in valleys or on streambanks
near rocks, in the Namakwaland
Klipkoppe, the Bokkeveld Mountains
and extending from the Cederberg to
Bredasdorp in the Western Cape.
Flowers October to January.

RIAAN DE VILLIERS

GENTIANACEAE **341**

CARISSA
There are four species of *Carissa* in southern Africa of which one occurs in Namaqualand.

Carissa bispinosa
noem-noem
APOCYNACEAE

An evergreen, densely branched, spiny shrub up to 2m tall, with greenish bark and milky sap. Spines are forked, sometimes very small or absent and can be slender or stout and woody. The elliptic, glossy leaves are dark green above and paler below. The fragrant, trumpet-shaped flowers are white. The fleshy, edible fruits are red to black when ripe.

Found in rocky areas in the mountains of Namaqualand, extending southwestwards to the mountains of the Western and Eastern Cape and northwards into Namibia.
Flowers June to August.

PACHYPODIUM
There are five species of *Pachypodium* in southern Africa of which one occurs in Namaqualand.

Pachypodium namaquanum
elephant's trunk, halfmens
APOCYNACEAE

An erect, deciduous, single-stemmed, rarely branched, spiny succulent up to 3m tall. Cylindrical stem, 10–30cm in diameter, tapering at the top, covered with pairs of long, straight, spreading, persistent spines; tops of the stems tilted to the north at about a 30° angle. Leaves sword-shaped, crowded at the top of the stem; densely velvety with star-shaped hairs; margin undulate. Flowers yellow-green, with deep red-purple throat; clustered at the top of the stem.

Found on extremely dry, rocky slopes in the Richtersveld, extending eastwards into northern Bushmanland and northwards into Namibia.
Flowers August to November.

HOODIA
There are 12 species of *Hoodia* in southern Africa of
which two occur in Namaqualand.

Hoodia gordonii
ghobba, muishondghaap, jakkalsghaap
APOCYNACEAE

An erect, spiny succulent up to 50cm high.
The leafless stems are 2.5–5cm in diameter,
with 11–17 angles, and bear spines
6–12mm long. Flesh-pink, saucer-shaped
flowers, up to 10cm in diameter, are thin
and papery and emit a putrid odour. Eaten
by stock and wild animals in times of
drought. The young growth is also eaten
by humans and is often pickled.

Found on rocky slopes or flats in the
Richtersveld, Namakwaland Klipkoppe
and also elsewhere in the Northern Cape,
Western Cape, Free State and Namibia.
Flowers January to May.

KOBUS MÜLLER

LARRYLEACHIA
There are five species of *Larryleachia* in southern
Africa of which three occur in Namaqualand.

Larryleachia cactiformis
hondebal
APOCYNACEAE

A succulent up to 15cm high, with 1–10
stems. The whitish-green, hairless and
leafless stems are 3–6cm in diameter,
with 12–16 angles. The bell-shaped
flowers, 6–15mm in diameter, are white to
yellow, barred with purple-red to almost
uniformly purple- or wine-red, and emit a
putrid scent.

Found on rocky slopes, often on
northern aspects, in the Richtersveld,
Namakwaland Klipkoppe and western
Bushmanland.
Flowers February to April.

ZELDA WAHL

ANNELISE LE ROUX

QUAQUA
There are 19 species of *Quaqua* in southern Africa of which 13 occur in Namaqualand.

Quaqua acutiloba
aroena
APOCYNACEAE

A clump-forming succulent up to 12cm high. The leafless stems, branching from the base, have 4 or 5 angles and are up to 2cm in diameter; tubercles on the stems end in sharp points. Flowers are borne in clusters of 1–3 in the grooves of the stem, opening in close succession; the inside is blotched blackish or purple-black on a yellow background, rarely uniformly yellow.

Found in firm, sandy soil on flats in Namaqualand, extending southeastwards into the Western Cape and northwards into Namibia.
Flowers April to August.

NICK HELME

Quaqua framesii
aroena
APOCYNACEAE   endemic

A clump-forming succulent up to 40cm high. The leafless stems, branching from the base, have 4 or 5 angles and are up to 2cm in diameter; tubercles on the stems are irregularly arranged and end in large, hard, yellow-tipped teeth. Flowers, yellow-green on the outside and bright yellow on the inside, are borne in clusters of 2–10 and open simultaneously.

Found in firm, sandy soil on flats among bushes on the Knersvlakte.
Flowers April to July.

344 APOCYNACEAE

Quaqua incarnata
aroena
APOCYNACEAE

A clump-forming succulent up to 30cm
high. The leafless stems, branching from
the base, have 4 angles and are up to
2.5cm in diameter; tubercles on the stems
end in stout, conical teeth. Flowers are
borne in clusters of 4–10; they are pale
pink to whitish on the outside and white,
cream-coloured or pale yellow on the
inside and are hairless except for stiff hairs
mainly in the mouth of the flower tube
and on the base of the lobes.

Found on rocky slopes and loamy flats
in Namaqualand, extending southwards to
Bokbaai and northwards into Namibia.
Flowers May to September.

ZELDA WAHL

Quaqua mammillaris
aroena
APOCYNACEAE

A clump-forming succulent up to 40cm
high. The leafless stems, branching from
the base and higher up, have 4 or 5 angles
and are up to 4cm in diameter; tubercles
on the stems are stout but sharply
pointed. Pale yellow flowers, dotted with
purple-black or uniformly dark purple-
black, open simultaneously and are borne
in clusters of 3–15; the petals are covered
on the inside with small papillae each
topped with a hair. When in full flower
the plant has a strong putrid scent.

Found on rocky slopes or sandy flats in
Namaqualand, extending southeastwards
to the Little Karoo and northwards into
Namibia.
Flowers May to July.

ZELDA WAHL

APOCYNACEAE **345**

JEAN SMITH

Stapelia acuminata
aasblom

APOCYNACEAE   endemic

A clump-forming succulent up to 12cm high. The leafless stems, branching from the base, have 4 angles; the tubercles on the stems are sharply pointed. The flowers are pale red-brown to yellow or purplish with fine white to cream-coloured transverse lines, finely wrinkled; the petals are sharply pointed and are usually hairy along the margin.

Found on dry rocky slopes to flat areas, growing under bushes, in the Namakwaland Klipkoppe and on the Knersvlakte.
Flowers April to June.

PIETER VAN WYK

Stapelia hirsuta var. *gariepensis*
aasblom

APOCYNACEAE

A loosely clumped succulent up to 12cm high. The leafless stems are often decumbent-erect and have 4 angles; the tubercles on the stems are blunt. The flowers, mainly red-purple on the inside, sometimes with buff to yellowish marks, are borne on stalks 2–7cm long and lie on the ground; the petals are covered with long, soft, pale purple hairs that are densest where the lobes join. When in flower, the plant emits a putrid scent.

Found on gentle to steep, rocky slopes in the Richtersveld and in Namibia.
Flowers May to September.

Stapelia hirsuta var. *hirsuta*
aasblom

APOCYNACEAE

A succulent up to 20cm high. The leafless
stems, 1–1.7cm in diameter, often form
large mats and have 4 angles; the
tubercles on the stems have small, soft
spines when young but become blunt
with age. Flowers are borne near the base
of the stems and lie on the ground; the
petals are dark purple with cream on the
wrinkles inside and are recurved; there is a
thick cushion of hairs in the centre.

    Found mainly among bushes on
lower, rocky slopes and flat areas in
Namaqualand, extending into the Western
and Eastern Cape.
Flowers May to July.

PIETER VAN WYK

Stapelia similis
aasblom

APOCYNACEAE

A succulent up to 12cm high. The leafless
stems, up to 1cm in diameter, are closely
packed into tight clumps and have
4 angles; the tubercles on the stems
have small, soft spines when young but
become blunt with age. Yellow-brown to
black-brown flowers, deeply transversely
wrinkled, are borne on stalks 3–5.5cm long
and lie flat on the ground.

    Found on rocky to gravelly slopes
under small bushes or next to stones in the
Richtersveld and Namibia.
Flowers May to July.

PIETER VAN WYK

JEAN SMITH

DUVALIA
There are 10 species of *Duvalia* in southern Africa of which one occurs in Namaqualand.

Duvalia caespitosa
APOCYNACEAE

A clump-forming succulent up to 5cm high. The leafless stems, branching from the base, are up to 2cm in diameter and have 4 or 5 angles; the tubercles on the stems are sharply pointed when young but become blunt with age. Red- to purple-brown flowers are borne in clusters of 1–3 and open in succession; the petals are conspicuously hairy in the centre. When in full flower the plant has a putrid scent.

Found on rocky slopes under bushes in relatively moist areas in Namaqualand, extending southeastwards into the Western and Eastern Cape and northwards into Namibia. **Flowers March to October.**

PIETER VAN WYK

HUERNIA
There are about 30 species of *Huernia* in southern Africa of which three occur in Namaqualand.

Huernia namaquensis
kopseer

APOCYNACEAE   endemic

A clump-forming succulent up to 5cm high. The leafless stems, branching from the base, are up to 2cm in diameter and have 5 angles; the tubercles on the stems are sharply pointed. The flowers, facing horizontally or nodding, are pinkish-cream on the outside and whitish-cream to pale yellow, usually spotted with reddish or purple, on the inside.

Found on stony quartzite or quartz gravel, usually in crevices, in the Richtersveld and Namakwaland Klipkoppe, extending northwards into Namibia. **Flowers May to October.**

Tromotriche pedunculata

APOCYNACEAE

A clump-forming succulent up to 10cm
high. The leafless stems, branching from
the base, are up to 1.5cm in diameter
and have 4 angles; the tubercles on the
stems are blunt. The flowers, borne on
stalks 5–20cm long, are pale green on the
outside and smooth to wrinkled, varying
in colour from olive-green or yellow to
purple-brown on the inside; they have a
sharp urine odour.

Found on slopes with white, quartz
pebbles and firm, red sand, under bushes,
in Namaqualand, extending southwards to
Touwsrivier and northwards into Namibia.
Flowers May to June.

MARYNA KOHRS

Orbea namaquensis

APOCYNACEAE   endemic

A clump-forming succulent up to 8cm
high. The leafless stems, branching from
the base, are up to 2.5cm in diameter
and have 4 angles; the tubercles on the
stems are sharply pointed. Yellow flowers,
spotted and lined with purple-brown, are
borne at ground level; they have a sharp
urine odour.

Found on sandy plains under bushes
in the Richtersveld and Namakwaland
Klipkoppe.
Flowers March to August.

ANDRE VAN WYK

GOMPHOCARPUS
There are eight species of *Gomphocarpus* in southern Africa of which three occur in Namaqualand.

Gomphocarpus cancellatus
dermhout, bergtonteldoosbos
APOCYNACEAE

A branched shrub up to 50cm high. The elliptic-oblong, leathery leaves, up to 5cm long and 3.3cm across, are whitish below with green veins. White flowers are borne in clusters of 12–30. When the inflated fruits open, the long, silky, white hairs attached to the seeds burst out and the seeds are swept away by the wind. The root has been used as a medicine for stomach ailments.

Found on rocky slopes in the Richtersveld and Namakwaland Klipkoppe, extending into mountainous areas of the Western and Eastern Cape.
Flowers May to September.

Gomphocarpus fruticosus
tonteldoosbos
APOCYNACEAE

A shrub with slender, erect branches up to 1.5m high. The yellow-green leaves, up to 15cm long and 1.5cm across, are narrowly sword-shaped and pointed at the tip. Cream-coloured to yellow flowers are borne in clusters of 6–10. Fruits are inflated and hairy.

Found in sandy soil on plains, often in dry riverbeds and along roadsides, in Namaqualand and throughout South Africa and Namibia.
Flowers July to September.

ASTEPHANUS

There are two species of *Astephanus* in southern Africa of which one occurs in Namaqualand.

Astephanus triflorus

APOCYNACEAE

A perennial herb up to 1m high, with slender stems 1mm in diameter twining into itself or into other bushes. The grey-green, leathery leaves, up to 2cm long and 5mm across, are minutely hairy. The 2–9 bell-shaped flowers, whitish on the inside and pinkish to whitish on the outside, are borne in sparse clusters along the stem; the flower tube has 5 hairy patches on the inside.

Found on cool, south-facing slopes in rocky areas among shrubs in Namaqualand, extending southeastwards to George.
Flowers June to September.

ANNELISE LE ROUX

SARCOSTEMMA

There are two species of *Sarcostemma* in southern Africa of which one occurs in Namaqualand.

Sarcostemma viminale

spantou-melkbos

APOCYNACEAE

A flat, trailing or spreading shrub up to 50cm high but up to 3m in diameter; the fleshy, blue-green stems contain milky sap. The scale-like, hairy leaves, pressed close to the stems, are shed quickly, giving the impression that the plant is leafless. Sweetly scented yellow flowers up to 1.5cm in diameter are borne in clusters. The 2-horned fruits split open to produce white, fluffy seeds that easily become airborne.

Found in the Richtersveld, Namakwaland Klipkoppe, on the Knersvlakte and also widespread elsewhere in Africa, Arabia, India, Nepal and Australia.
Flowers March to December.

MARYNA KOHRS

APOCYNACEAE **351**

ANNELISE LE ROUX

MICROLOMA
There are about 10 species of *Microloma* in southern Africa of which seven occur in Namaqualand.

Microloma calycinum
bokhorinkie
APOCYNACEAE

A slender, twining shrub up to 80cm high, within other plants. The slightly fleshy, linear to sword-shaped leaves, up to 2cm long and 5mm across, bend downwards. Bright pink-red flowers are borne in clusters; the stamens form a central column. A very palatable plant to stock.

Found in rocky hills in Namaqualand and also in Namibia.
Flowers June to September.

PIETER VAN WYK

Microloma incanum
bokhorinkie
APOCYNACEAE

A shrub up to 80cm high. Grey-green stems, often twining into one another, are densely covered with small, white, adpressed hairs. The grey-green, linear leaves, up to 3.5cm long and 7mm across, are thinly to thickly hairy; the margin is recurved. The flowers, pink in the basal half and whitish in the upper half, are borne in clusters; the stamens form a central column. The plants have a strong smell of aloes when broken and seem to be left alone by stock and wild animals.

Found on dry slopes and flat areas in the Richtersveld and Namakwaland Klipkoppe, extending eastwards to Carnarvon and northwards into Namibia.
Flowers July to September.

Microloma namaquense
bokhorinkie

APOCYNACEAE   endemic

A slender, twining perennial herb. The linear, leathery leaves are opposite, up to 3.5cm long and 2mm across; the margin is strongly recurved. Bright red, urn-shaped flowers about 6mm long are borne in a loose cluster of 3–6; the petals overlap to form a spiral rosette. A very palatable plant to stock.

Found in mountainous areas in the Namakwaland Klipkoppe.
Flowers June to August.

VERONICA ESTERHUIZEN

Microloma sagittatum
bokhorinkie

APOCYNACEAE

A slender, twining perennial herb. The grey-green, arrow-shaped leaves, up to 3cm long and 7mm across, are opposite and covered in minute hairs; the margin is recurved. White to deep pink flowers about 1cm long are borne in clusters of 3–10; the petals hardly open. A very palatable plant to stock.

Found on stony flats or slopes throughout Namaqualand, extending into the Western Cape.
Flowers June to September.

ANNELISE LE ROUX

APOCYNACEAE **353**

ANNELISE LE ROUX

HELIOTROPIUM
There are about 15 species of *Heliotropium* in southern Africa of which four occur in Namaqualand.

Heliotropium curassavicum
BORAGINACEAE   alien

A spreading annual or perennial herb up to 35cm high. The fleshy, oblong, blunt, blue-green leaves taper downwards into a short stalk. White flowers are borne at the end of a recurved stalk; the stamens arise from about halfway up the flower tube. A weed originally from the Americas.

Found in sandy, damp, saline soil, often in disturbed areas, in Namaqualand and other dry areas of southern Africa.

ANNELISE LE ROUX

Heliotropium tubulosum
BORAGINACEAE

A very hairy perennial herb up to 35cm high. The sword-shaped, pale grey-green leaves are densely hairy and pointed. White flowers are borne at the end of a recurved stalk; the flower tube is slightly longer than the calyx and the stamens arise from about halfway up the flower tube.

Found in sandy soil in dry riverbeds or among rocks, in the Richtersveld, on the Knersvlakte and also in Namibia.
Flowers April to October.

Codon royenii
soetdoringbos, suikerkelk
BORAGINACEAE

An erect perennial shrub up to 1m high and covered all over with straight, white spines. The egg-shaped leaves are deeply undulate and have long, white spines. The white flowers have purplish-red or greenish stripes and are about 3cm long; the lobes are much shorter than the tube. Children enjoy sucking the flowers for the sweet nectar.

Found on rocky slopes and in dry riverbeds in Namaqualand, extending eastwards into Bushmanland and the Tankwa Karoo and northwards into Namibia.
Flowers May to October.

ZELDA WAHL

Codon schenckii
BORAGINACEAE

An erect perennial herb up to 75cm high and covered all over with straight, white spines. The egg-shaped leaves are deeply undulate and have long, white spines. The yellow flowers are shorter than 2cm; the lobes are only slightly shorter than the tube.

Found in sandy watercourses and on rocky slopes in the Richtersveld and Namibia.
Flowers in September.

NORBERT JÜRGENS

BORAGINACEAE **355**

HEATHER BURGER

Lobostemon echioides

BORAGINACEAE

A shrub up to 80cm high. Branches are covered with a mixture of fine and coarse hairs. The stalkless, egg-shaped leaves are covered with soft, silvery hairs. The blue flowers, hairy on the outside and with petals spread horizontally, are borne in clusters at the ends of branches; the stamens are the same length and are exserted above the petals; the style is smooth.

Found in sandy soil on rocky flats and mountain slopes in the Richtersveld and Namakwaland Klipkoppe, extending into the Western and Eastern Cape. **Flowers August to October.**

ANNELISE LE ROUX

Lobostemon fruticosus

agtdaegeneesbos
BORAGINACEAE

A shrub up to 80cm high. Branches are densely hairy, with a mixture of fine and coarse hairs. The inversely sword-shaped to inversely egg-shaped leaves have a prominent midrib on the lower surface and both surfaces are covered with adpressed hairs. The blue, pink or rarely white flowers are funnel-shaped; the stamens are of three lengths, all more or less shorter than the flower tube; the style is hairy.

Found in sandy soil on rocky slopes in the Namakwaland Klipkoppe, extending southwards to Worcester. **Flowers August to October.**

Lobostemon glaucophyllus
BORAGINACEAE

A shrub up to 1m high. Branches are hairless. The grey-green, narrowly oblong to sword-shaped leaves are slightly fleshy; the margin has stiff white bulbous-based hairs adpressed along the margin. The blue, rarely pink, flowers are funnel-shaped; the stamens are of three lengths, longer than, or just included in the flower tube, all except the short stamen fused to the tube; the style is hairy.

Found in sandy, flat areas on rocky mountain slopes in the Namakwaland Klipkoppe, extending southeastwards to Laingsburg.
Flowers July to November.

ANDRE VAN WYK

ANCHUSA
There are three species of *Anchusa* in southern Africa of which one occurs in Namaqualand.

Anchusa capensis
koringblom, vergeet-my-nietjie
BORAGINACEAE

An erect perennial herb up to 75cm high. The erect, grooved stems are hairy and usually unbranched. The narrowly egg-shaped leaves are densely hairy above and below. The flowers are about 1cm in diameter, changing from purple to blue.

Found in sand in the Namakwaland Klipkoppe, the rest of South Africa and in Namibia.
Flowers September to November.

ZELDA WAHL

Trichodesma africanum
brandnetel
BORAGINACEAE

An annual or perennial herb up to 50cm high. The erect stems have stiff, short, white hairs. The opposite, egg-shaped to oblong leaves have stiff, short, white hairs. The pale pink flowers have a yellow centre with 5 brownish spots.

Found in dry watercourses or on rocky slopes in Namaqualand and other dry areas in Africa.
Flowers June to October.

ANNELISE LE ROUX

AMSINCKIA
There are two alien species of *Amsinckia* in southern Africa and both are found in Namaqualand.

Amsinckia retrorsa
neukbos, kakiebos, ystergras
BORAGINACEAE   alien

An erect annual herb up to 30cm high. The leaves are 10cm long, linear to narrowly elliptic, and both sides have stiff, white hairs. Yellow flowers about 1cm in diameter are borne on one side of the curled stalk; the hairy calyx is 5–7mm long. A weed introduced from the western United States.

Usually found in disturbed areas in Namaqualand and in the rest of the winter-rainfall area of the Western Cape.
Flowers August to October.

ZELDA WAHL

CONVOLVULUS
There are about 15 species of *Convolvulus* in southern
Africa of which two occur in Namaqualand.

Convolvulus capensis
Cape bindweed, trompetblom
CONVOLVULACEAE

A creeping or twining perennial herb. The
deeply lobed leaves are up to 3.5cm long
and thinly to densely hairy. Funnel-shaped,
white, pale pink or mauve flowers about
3cm in diameter are borne singly on long
leafless stalks.

Usually found along roadsides in flat
sandy areas in the Namakwaland Klipkoppe
and on the Knersvlakte, extending into the
Western and Eastern Cape.
Flowers August to November.

RIAAN DE VILLIERS

Convolvulus sagittatus
wild bindweed, wildetrompetblom
CONVOLVULACEAE

A perennial herb with stems up to 1m
long that lie on the ground. The deeply
lobed leaves, up to 5cm long, are 3 times
as long as broad, thinly to densely hairy.
Funnel-shaped, pale pink or mauve
flowers about 3.5cm in diameter are borne
singly on long leafless stalks.

Found on rocky flats and slopes
in the Namakwaland Klipkoppe and
widespread in southern Africa, tropical
Africa and Arabia.
Flowers September to December.

HEATHER BURGER

CONVOLVULACEAE **359**

ANNELISE LE ROUX

NICOTIANA
There are about 70 species of *Nicotiana* worldwide of which two alien species occur in Namaqualand.

Nicotiana glauca
tobacco tree, Jan-twak

SOLANACEAE   alien invader

An erect, loosely branched shrub or small tree up to 3m tall. The egg-shaped, leathery, greyish-green leaves are up to 10cm long. Yellow flowers up to 4cm long are borne in loose clusters at the ends of branches. Introduced from South America and now a widespread weed. The wood is poisonous to humans and animals.

Found along riverbeds, roadsides and in other disturbed ground throughout Namaqualand and other dry areas of South Africa.
Flowers January to December.

HEATHER BURGER

LYCIUM
There are about 20 species of *Lycium* in southern Africa of which nine occur in Namaqualand.

Lycium amoenum
(previously misidentified as *Lycium ferocissimum*)
karriedoring

SOLANACEAE

A spreading shrub up to 1.5m high. Stems have sturdy thorns, set at right angles and about 3–5cm long. Leaves slightly fleshy, hairless, spoon-shaped and borne in clusters along the stem and on the thorns. Flowers purple, bell- to funnel-shaped, up to 1.5cm long; lobes longer than flower tube by a third or more; style and stamens longer than the flower tube. Fruits red, round and up to 1cm in diameter.

Found along watercourses and on rocky slopes in Namaqualand, extending into the Western Cape, the Free State and Lesotho and into Namibia.
Flowers August to September.

360 SOLANACEAE

Lycium bosciifolium
(previously misidentified as *Lycium oxycarpum*)
brosdoring
SOLANACEAE

A lax shrub up to 2m tall, with thin spines about 1cm long. The slightly fleshy, hairless, narrowly elliptic leaves, borne in clusters, are 1–1.5cm long and 2–5mm across. The long, funnel-shaped flowers, cream-coloured or pale purple with dark purple veins, are 1.5–2cm long; the lobes are less than a third as long as the flower tube; the style and stigmas are longer than the flower tube. The yellow or red fruits are elliptic, up to 6mm long.

Found in sandy, flat areas in Namaqualand, Bushmanland, the Western and Eastern Cape and in Namibia. **Flowers August to September.**

ANNELISE LE ROUX

Lycium cinereum
kriedoring
SOLANACEAE

A much-branched, spiny shrub up to 1m high, with thin spines about 1cm long on young branches, later developing into thick spines about 35cm long, with many clusters of leaves. The slightly fleshy, linear to narrowly elliptic, hairless leaves, borne in clusters on the stems or thorns, are up to 1.5cm long and 2mm across. The trumpet-shaped flowers, white or pale purple with dark purple veins, are less than 1cm long; the lobes are less than a third as long as the flower tube; the style and stigmas are longer than the flower tube. The yellow or red fruits are elliptic, up to 7mm long and 6mm across.

Found in sandy, flat areas or on rocky slopes in Namaqualand and also in the rest of South Africa, Lesotho and Namibia. **Flowers August to September.**

ANNELISE LE ROUX

SOLANUM
There are about 50 species of *Solanum* in southern Africa of which seven occur in Namaqualand.

Solanum burchellii
lemoenbossie, tandpynbos
SOLANACEAE

A much-branched, spiny shrub up to 50cm high, with spines up to 1cm long on the branches. The yellow-green, hairy leaves are shallowly lobed, with scattered spines along the midrib. The flowers are purple and about 1.5cm in diameter. The round fruits are bright orange when ripe and about 1cm in diameter. The orange berries are reputed to cure toothache.

Found on rocky slopes in the Richtersveld and Namakwaland Klipkoppe, extending eastwards into the Northern Cape and northwards into Namibia. **Flowers March to October.**

Solanum giftbergense
gifappeltjie
SOLANACEAE

A much-branched shrub up to 50cm high, with numerous spines up to 1cm long on the stems. The bright green leaves are lobed, with spines along the stalk and veins. The flowers are purple and about 2cm in diameter. The round fruits are orange when ripe and about 1cm in diameter.

Found in dry watercourses and rocky areas in Namaqualand, extending southwards to Clanwilliam and northwards into Namibia. **Flowers April to September.**

Solanum guineense
SOLANACEAE

A much-branched shrub up to 1m high,
without spines. The bright green, egg-
shaped leaves are softly leathery. The
flowers are pale mauve and about 2.5cm
in diameter. The round fruits are red when
ripe and about 1.5cm in diameter.

Found in sandy soil in dry
watercourses, on flats and in rocky areas
in Namaqualand, extending into the
Western and Eastern Cape.
Flowers April to August.

VERONICA ESTERHUIZEN

Solanum nigrum
nightshade, nastergal
SOLANACEAE   alien

An erect annual herb up to 50cm high,
with angular and hairy stems. The soft,
egg- or sword-shaped leaves are up to
10cm long and more or less hairy on both
surfaces; the margin is entire or toothed.
White flowers up to 5mm in diameter
are borne in clusters of 5–10; the stamens
are of equal length and the filaments are
short. The round fruits are black when
ripe and about 1cm in diameter. A weed
originally from Eurasia.

Found in disturbed areas in Namaqualand and in southern and tropical Africa.
Flowers January to December.

ANNELISE LE ROUX

ANNELISE LE ROUX

DATURA
There are four alien species of *Datura* in southern
Africa and all occur in Namaqualand.

Datura inoxia
devil's trumpet, stinkblaar
SOLANACEAE   alien

A spreading annual herb up to 1m
high. Stems and leaves are covered with
short, soft, greyish hairs that give the
whole plant a greyish appearance. The
alternate, elliptic leaves are 3–7cm long.
The white, trumpet-shaped flowers
have green veins; the stigma is exserted
beyond the anthers; the flowers emit a
fragrant scent at night. Fruits are ovoid,
with numerous slender spines, and are
about 5cm in diameter. A weed originally
from Central and South America.

Found in dry watercourses and
along roadsides in Namaqualand and
throughout the world.
Flowers March to December.

ANNELISE LE ROUX

MONTINIA
There is one species of *Montinia* in southern Africa
and in Namaqualand.

Montinia caryophyllacea
peperbos, t'iena
MONTINIACEAE

An erect shrub up to 1.5m high. The
bluish-green leaves, up to 2.5cm long, are
simple, egg- to sword-shaped and entire.
White flowers with a green centre are
borne in loose clusters towards the ends
of branches; male flowers, about 5mm in
diameter, and female flowers, about 8mm
in diameter, are borne on separate plants.
Female plants produce oblong green
fruits. The stem is very hard and was used
by the San for digging.

Found on rocky slopes in Namaqualand,
extending into the Western and Eastern
Cape, Namibia and Angola.
Flowers June to November.

Menodora juncea
blombiesie

OLEACEAE

A stiffly erect shrub up to 1.5m high,
with silvery-hairy branches. The reduced,
scattered, linear leaves are pressed closely
to the stem and are inconspicuous. Yellow
flowers up to 2.5cm in diameter are
borne in clusters of 1–3; there are only 2
stamens. The fleshy, round fruits become
dark purple when mature.

Found in sandy valleys and
watercourses in the Richtersveld and
Namakwaland Klipkoppe and in the Great
Karoo and Little Karoo.
Flowers September to October.

VERONICA ESTERHUIZEN & ANNELISE LE ROUX

Olea europaea
subsp. *africana*
wild olive, wilde-olienhout

OLEACEAE

A small to medium-sized tree up to 5m
tall. The narrow, oblong-elliptic leaves are
dark green above and paler below, where
they are densely covered with small,
silvery, golden or pale green scales. Many
small, white flowers are borne on a long
stalk. The round fruits are green to red to
black when ripe. Indigenous people use an
extract from the leaves boiled in water as
coffee. The tree is also good for timber.

Found along streambanks, in mountain
kloofs and on rocky ledges in the Namakwaland Klipkoppe. It is also widespread
throughout the rest of South Africa.
Flowers October to March.

ANNELISE LE ROUX

OLEACEAE **365**

NOEL VAN ROOYEN

COLPIAS
There is only one species of *Colpias* in southern Africa and in Namaqualand.

Colpias mollis
klipblom
SCROPHULARIACEAE

A much-branched, cushion-forming shrub up to 10cm high, with softly hairy, brittle branches. The egg-shaped, stalked leaves are 2–4cm long and softly hairy; the margin is toothed. Scented, cream-coloured to bright yellow flowers with 5 lobes, a short tube and 2 short pouches are borne on slender stalks.

Found in crevices of large, granitic rocks, usually facing south, in the Richtersveld, Namakwaland Klipkoppe and also in the Bokkeveld Mountains. **Flowers July to September.**

JEAN SMITH

HEMIMERIS
There are six species of *Hemimeris* in southern Africa of which three occur in Namaqualand.

Hemimeris racemosa
bobbejaangesiggies
SCROPHULARIACEAE

A branched or unbranched annual herb up to 40cm high, with hairy stems. The elliptic to egg-shaped leaves, up to 5cm long and 7cm across, are minutely hairy; the margin is toothed or occasionally entire. Yellow flowers, about 1cm in diameter and 2-lipped, with 2 orange or maroon spots on the upper lip, are borne in groups on slender stalks; the corolla has a well-developed basal pouch.

Found in moist places, often in the shade of large boulders, throughout Namaqualand and the winter-rainfall area. **Flowers August to September.**

There are about 70 species of *Diascia* in southern Africa of which 14 occur in Namaqualand.

Diascia namaquensis
bokhorinkies, ramhorinkies
SCROPHULARIACEAE   endemic

An annual herb up to 40cm high, branching from the base. The leaves are inversely sword-shaped, up to 5cm long and 1.2cm across, toothed to shallowly lobed. Salmon to purple-red flowers, about 2cm in diameter and with 2 diverging spurs 0.5–2cm long, are borne singly on long stalks; the stamens project forwards and the posterior filaments are thickened or bent slightly backwards about halfway along their length. Fruits are long and narrow.

Found in sandy soil throughout Namaqualand and on the Bokkeveld Escarpment.
Flowers August to September.

ZELDA WAHL

Diascia rudolphii
SCROPHULARIACEAE   endemic

An erect annual herb up to 6cm high. The basal leaves, 2–6cm long and 1–1.5cm across, are egg-shaped to elliptic; the margin is toothed to scalloped. The pale orange flowers have a yellow centre and are about 1–2cm in diameter; the filaments have dense, dark purple or white hairs.

Found in sandy soil in the Richtersveld, Namakwaland Klipkoppe and on the Knersvlakte.
Flowers July to September.

HELGA VAN DER MERWE

ZELDA WAHL

There are about 60 species of *Nemesia* in southern Africa of which 18 occur in Namaqualand.

Nemesia bicornis
kappieblommetjie
SCROPHULARIACEAE

An annual herb up to 60cm high. The leaves are narrow, sword-shaped and toothed. White to pale lilac flowers, with purple veining and 2–4 orange to yellow, hairy protuberances, are borne sparsely on long stalks; the spur is 4–5mm long, projecting backwards and then deflecting downwards in the distal half. Fruits are 2-horned, broader than long, and the base of the capsule is slightly oblique.

Found on coastal and inland flats throughout Namaqualand, extending southwards towards Agulhas.
Flowers August to September.

RIAAN DE VILLIERS

Nemesia cheiranthus
kappieblommetjie
SCROPHULARIACEAE

An annual herb up to 40cm high. The leaves are elliptic to sword-shaped and slightly toothed. Flowers about 1.5–2.5cm long are borne sparsely on long stalks; the upper lip with linear lobes is about twice the length of the lower lip and is white or lavender with a rectangular orange spot at the base, often surrounded by blue and marked with purple; the lower lip, with broader lobes, is yellow or white, with 2, smooth, convex, orange protuberances at the base; the spur is 3–5mm long. Fruits are 2-horned, longer than broad, and the base of the capsule is slightly oblique.

Found in loamy sand on slopes and flats in the Namakwaland Klipkoppe, the Bokkeveld Mountains and the Cederberg.
Flowers August to September.

Nemesia ligulata
kappieblommetjie
SCROPHULARIACEAE

An annual herb up to 30cm high. The
leaves are sword-shaped and very slightly
toothed. Flowers about 1.5–1.8cm long
are borne sparsely on long stalks; the
upper corolla lip is white or pale yellow,
with linear to narrowly oblong lobes and
a rectangular yellow to orange spot at
the base; the lower lip is yellow or orange
with 2 smooth, convex protuberances
at the base; the spur is 5–7mm long,
projecting downwards and forwards.
Fruits are 2-horned, as long as broad, and
the base of the capsule is slightly oblique.

Found on sandy slopes and flats in
the Richtersveld and Namakwaland
Klipkoppe, extending further southwards
into the mountains of the Karoo in the
Western Cape.
Flowers August to October.

VERONICA ESTERHUIZEN

Nemesia versicolor
kappieblommetjie, weeskindertjies
SCROPHULARIACEAE   endemic

An annual herb up to 30cm high. The
leaves are narrow, sword-shaped and
toothed to entire. Flowers about 1.5–2cm
long are borne sparsely on long stalks;
the upper corolla lip is yellow or pale
blue-violet, with a white or yellow patch
and darker lines at the base; the lower
lip, with 2 raised, diverging, convex
protuberances forming a V at the base, is
covered with yellow or orange hairs; the
spur is 6–8mm long, projecting backwards
and downwards in the distal third. Fruits
are wider than long and slightly oblique at
the base.

Found on sandy flats in the Namakwaland Klipkoppe.
Flowers July to September.

HEATHER BURGER & HELGA VAN DER MERWE

ANNELISE LE ROUX

APTOSIMUM
There are about 20 species of *Aptosimum* in southern
Africa of which four occur in Namaqualand.

Aptosimum indivisum
viooltjie
SCROPHULARIACEAE

A small, rounded, compact shrub up
to 5cm high. The numerous leaves are
narrowly elliptic and entire, with stalks
as long as the blades. The stalkless, blue
to violet flowers, about 2.5cm long, have
darker patches at the base and a whitish
throat. A palatable plant to stock.

Found on stony flats in the Richtersveld
and on the Namakwaland Klipkoppe,
extending southeastwards to the inland
mountains of the Western Cape.
Flowers August to November.

ZELDA WAHL

Aptosimum spinescens
doringviooltjie, aambeibos
SCROPHULARIACEAE

A spreading shrub up to 30cm high, with
spines up to 1.5cm long on the stem. The
rigid leaves, up to 2.5cm long, are narrow,
linear and entire, becoming spiny with
age. The flowers, about 1.5cm in diameter,
have purple, dark blue or occasionally
white petals with a fawn tube. A palatable
plant to stock.

Found in sandy, flat areas in the
Namakwaland Klipkoppe, Bushmanland,
the Western and Eastern Cape, Free State
and southern Namibia.
Flowers August to November.

There are seven species of *Peliostomum* in southern Africa of which two occur in Namaqualand.

Peliostomum leucorrhizum
viooltjie
SCROPHULARIACEAE

An erect to spreading shrub up to 30cm high, with pale stems near the base. The leaves, up to 1.9cm long, are linear to narrowly inversely egg-shaped and entire. The flower tube is funnel-shaped, pale with dark longitudinal lines; the lobes are purplish with darker markings in the throat and the anthers have long hairs. A palatable plant to stock.

Found on sandy flats in the Richtersveld, Namakwaland Klipkoppe and also in the rest of South Africa and in southern Namibia.
Flowers August to September, but can also flower after rain in summer.

ZELDA WAHL

Peliostomum virgatum
viooltjie
SCROPHULARIACEAE

A spreading shrub up to 25cm high, sparsely branched from the base. The sticky leaves are up to 12mm long, egg-shaped to elliptic and entire. Purple or dark blue flowers, with a narrowly funnel-shaped, pale tube with darker lines, are borne along the erect to spreading branches; the lobes are violet; the anthers have short hairs. A palatable plant to stock.

Found on slopes and flats throughout Namaqualand, Bushmanland, the Bokke-veld Escarpment and the Cederberg.
Flowers August to November.

ZELDA WAHL

TEEDIA
There are two species of *Teedia* in southern Africa of which one occurs in Namaqualand.

Teedia lucida
klipkersie
SCROPHULARIACEAE

A pungent shrub about 0.5–1m high, with stems square in cross-section. The shiny, elliptic leaves, borne on winged stalks, are conspicuously finely toothed. The lilac or pink flowers are hairy at the throat of the flower tube and are marked with a star-shaped dark purple spot near the throat. The round fruits are black-purple or yellowish-brown.

Found on mountain slopes among rocks, often in rock crevices, in the Richtersveld, the Namakwaland Klipkoppe and also widespread in the rest of South Africa.
Flowers September to October.

JAMESBRITTENIA
There are about 80 species of *Jamesbrittenia* in southern Africa of which 14 occur in Namaqualand.

Jamesbrittenia amplexicaulis
SCROPHULARIACEAE   endemic

A sticky, foetid perennial herb up to 45cm high, with stems basally tufted from a woody rootstock. The broadly elliptic, upright leaves overlap and are glandular-hairy. Small, creamy-white to pale yellow flowers, streaked with brown at the base, are borne singly in upper leaf axils.

Found in dry watercourses in the Richtersveld and Namakwaland Klipkoppe.
Flowers July to October.

Jamesbrittenia fruticosa
SCROPHULARIACEAE

A branched, foetid shrub up to 1m high. The leaves, up to 3.5cm long, are elliptic, entire and sparsely hairy. Pale mauve flowers, up to 1.5cm in diameter and with dark purple markings, are borne in clusters at the ends of branches.

Usually found among rocks or in dry riverbeds throughout Namaqualand and in southern Namibia.
Flowers July to November.

ANNELISE LE ROUX

Jamesbrittenia glutinosa
SCROPHULARIACEAE

An aromatic annual or sometimes perennial herb up to 40cm high, covered with glandular hairs that are broad and flattened at the base. The egg-shaped leaves are toothed. The flowers, mauve, lilac, rose or white with 3 deep violet bars near the base of the lobes, are borne singly in leaf axils; the orange-yellow throat is laterally compressed.

Found in sandy or rocky watercourses and rocky mountain slopes in the Richtersveld and in southern Namibia.
Flowers July to September.

ANNELISE LE ROUX

VERA HOFFMANN

Jamesbrittenia pedunculosa
SCROPHULARIACEAE   endemic

A brittle annual or sometimes untidily branched perennial herb up to 40cm high. The egg-shaped or oblong leaves, 1–3cm long and 5–10mm across, are glandular-hairy and cold to the touch; the margin is lobed and toothed. The flowers are bright yellow, with a dark throat and two dark patches on the upper lip as well as one at the back of the inflated part of the slender tube; the tube widens into 5 lobes at the top; the upper lip is larger than the lower and slightly hooded at the base.

Found on rocky slopes, usually under large rocks in full shade, in the Namakwaland Klipkoppe.
Flowers August to September.

ZELDA WAHL

Jamesbrittenia racemosa
SCROPHULARIACEAE   endemic

An aromatic, erect annual herb up to 40cm high. The elliptic leaves, with glandular hairs, are up to 5.5cm long and irregularly toothed. White flowers with purple streaks and up to 1.5cm in diameter are borne in loose clusters at the ends of branches; the lobes are deeply notched.

Found in sandy soil throughout Namaqualand.
Flowers August to September.

374 SCROPHULARIACEAE

Jamesbrittenia ramosissima
SCROPHULARIACEAE

A dense, rounded, brittle, intricately branched shrub up to 1m high. Branches and leaves are densely covered with short, stalked, glistening glands and delicate glandular hairs. The alternate, broadly egg-shaped leaves are up to 1cm long; the margin usually has coarse, sharp teeth. The flowers, also thickly covered with glistening stalked glands and delicate glandular hairs, have a yellow centre and change from blue or mauve to white with age.

Found in rocky shelters, mostly on steep slopes but also in stony, dry riverbeds, north and south of the Orange River in the Richtersveld, extending eastwards to Augrabies Falls.
Flowers March to October.

TED OLIVER & LEONARD SEGAL

Lyperia tristis
evening flower, aandblom
SCROPHULARIACEAE

An erect, sticky annual herb up to 40cm high. The oblong to elliptic leaves are up to 6cm long, toothed and glandular-hairy. Cream-coloured, yellow, greenish or brownish flowers up to 1.5cm in diameter are borne in loose clusters at the top of the stem; they emit a strong scent at dusk, hence the common name.

Found in sandy soil on flats throughout Namaqualand, in dry areas of the Western Cape and in Namibia.
Flowers July to October.

ANNELISE LE ROUX

There are about 70 species of *Manulea* in southern Africa of which 18 occur in Namaqualand.

Manulea altissima
SCROPHULARIACEAE

An erect herb up to 60cm high. The inversely sword-shaped leaves, up to 8cm long, are mostly basal, have sticky hairs, and are narrowly and obscurely toothed. Scented white flowers with a yellow centre are borne in a cluster at the end of a long leafless stalk; the mouth of the flower tube is laterally compressed.

Found in sandy soil throughout Namaqualand, extending southwards towards Malmesbury in the Western Cape. **Flowers August to September.**

ALTA CHAMBERS

Manulea nervosa
SCROPHULARIACEAE

An annual herb up to 20cm high. The spathulate leaves, borne in a basal rosette, are up to 2cm long and taper to the leaf stalk; the margin is entire. White to mauve-blue flowers with a yellow centre are crowded together at the ends of the stalks.

Found on sandy flats along the eastern parts of the Namakwaland Klipkoppe and in Bushmanland. **Flowers June to September.**

HEATHER BURGER

Manulea silenoides

SCROPHULARIACEAE   endemic

An erect to spreading annual herb up to 10cm high, with glands or balloon-tipped hairs on the stems. The toothed, elliptic leaves are mostly in a basal rosette; both sides are hairy. The flowers, usually mauve to blue with a yellow and orange star-shaped patch in the centre, are borne in clusters at the end of the stalks; the 5 lobes are deeply divided into two and then shallowly divided again.

Found in sandy or stony soil, often on the slopes of granite hills, in the Namakwaland Klipkoppe.
Flowers August to September.

ZELDA WAHL

ZALUZIANSKYA
There are about 55 species of *Zaluzianskya* in southern Africa of which 11 occur in Namaqualand.

Zaluzianskya benthamiana

SCROPHULARIACEAE

An annual herb, branching from the base, up to 25cm high, with long, white hairs on the young branches. The hairy leaves, up to 2cm long, are linear to narrowly elliptic and entire or almost so. Pale yellow flowers with a deeper yellow centre are borne in clusters at the ends of branches; the hairy flower tube is about 1.4cm long and the petals are round.

Found in sandy areas throughout Namaqualand, the arid inland mountains from the Bokkeveld Mountains to Oudtshoorn and also in southern Namibia.
Flowers June to August.

ZELDA WAHL

SCROPHULARIACEAE **377**

ANNELISE LE ROUX

SELAGO
There are about 190 species of *Selago* in southern Africa of which 16 occur in Namaqualand.

Selago glabrata
t'gibbiebos, muishondbos
SCROPHULARIACEAE

A shrub with long slender branches, up to 35cm high and branching from the base. The hairless, linear, slightly fleshy leaves are tufted along the branches. White flowers are borne in small clusters on short stalks all along the branches.

Found on rocky slopes in Namaqualand, the Bokkeveld Escarpment and extending southwards to Worcester and the Little Karoo.
Flowers September to October.

RIAAN DE VILLIERS

Selago glutinosa
aarbossie
SCROPHULARIACEAE

A shrub up to 40cm high, with stems often appearing varnished. The slightly fleshy, needle-like leaves are not tufted as in most other *Selago* species. Intensely fragrant white flowers with a coarsely hairy calyx are borne close together at the ends of branches.

Found in rocky areas in the Namakwaland Klipkoppe, extending southwards to Montagu in the Western Cape.
Flowers September to October.

Selago namaquensis
(previously misidentified as *Selago albida*)
t'gibbiebos, muishondbos
SCROPHULARIACEAE   endemic

A shrub up to 30cm high, with tufted branches from a woody rootstock and hairy young branches. The hairy, elliptic to narrowly elliptic leaves are up to 5mm long; the margin is rolled slightly backwards. Numerous white, sometimes mauve flowers, about 5mm long and 3mm in diameter, are borne in round to oblong clusters.

Found in sandy soil throughout Namaqualand.
Flowers September to October.

DISCHISMA
There are 11 species of *Dischisma* in southern Africa of which five occur in Namaqualand.

Dischisma spicatum
SCROPHULARIACEAE

An erect, branched or unbranched, hairy annual herb, up to 30cm high. The narrow and spreading leaves, up to 4cm long and 4mm across, are entire to slightly toothed. White flowers up to 1.2cm long are borne close together on a long stalk at the ends of branches.

Found in sandy soil throughout Namaqualand, extending southwards to Piketberg in the Western Cape and also in Namibia.
Flowers August to September.

NICK HELME

HEBENSTRETIA
There are 24 species of *Hebenstretia* in southern
Africa of which 11 occur in Namaqualand.

Hebenstretia kamiesbergensis
SCROPHULARIACEAE   endemic

A shrub up to 60cm high, with erect, densely
leafy branches. The linear leaves are curved
backwards and profusely toothed. White
flowers with yellow spots are borne close
together at the ends of branches.

Found on the upper slopes of the
Kamiesberg.
Flowers September to December.

ZELDA WAHL

Hebenstretia parviflora
SCROPHULARIACEAE

An erect annual herb up to 25cm high,
branching from the base. The leaves are
up to 3cm long, linear and almost entire.
White flowers, with or without an orange
eye, are borne fairly closely set at the ends
of branches.

Found in sandy soil throughout
Namaqualand and also in dry areas of the
Northern, Western and Eastern Cape and
in Namibia.
Flowers July to October.

There are three species of *Rogeria* in southern Africa of which one occurs in Namaqualand.

Rogeria longiflora
PEDALIACEAE

An erect, slightly fleshy annual herb up to 1.5m high. The opposite, egg-shaped leaves are borne on outspread stalks 3–9cm long. The trumpet-shaped, white flowers have purple lines in the throat. The hard, spiny fruits have 2–8 conical spines and are 4–6cm long.

Found in sandy soil along drainage lines and on rocky foothills in the Richtersveld and in Namibia.
Flowers May to December.

ANNELISE LE ROUX & TED OLIVER

SESAMUM
There are about 10 species of *Sesamum* in southern Africa of which one occurs in Namaqualand.

Sesamum capense
PEDALIACEAE

An erect annual herb up to 1m tall. The lower leaves have 3–7 leaflets, oblong to linear; the upper leaves are smaller and have only 3 leaflets. Broadly trumpet-shaped, hairy, purple-violet flowers are borne singly in leaf axils. The woody fruits are cylindrical, with a sharp tip and basal knobs.

Found in loamy sand on flats in Namaqualand, in the eastern winter-rainfall region and also in central southern Africa.
Flowers at any time of the year after rain.

ANNELISE LE ROUX

PEDALIACEAE **381**

STACHYS
There are about 40 species of *Stachys* in southern Africa of which five occur in Namaqualand.

Stachys flavescens
LAMIACEAE

An erect, rigid, branched shrub up to 1m high, with yellowish, densely felted young branches and greyish older branches. The sword-shaped to oblong, pointed leaves are yellowish and densely felted on both sides. Yellow flowers are borne in several whorls at the ends of branches.

Found on rocky slopes in the Richtersveld and the Namakwaland Klipkoppe, the Bokkeveld Mountains and the Matsikammaberg.
Flowers September to November.

Stachys rugosa
vaaltee
LAMIACEAE

An erect, lax shrub up to 1.5m high, with white, velvety branches that darken to greyish-black with age. The linear to narrowly elliptic leaves are 2–6cm long and hairy on both surfaces; the veins are raised on the lighter coloured lower surface; the darker upper surface is slightly wrinkled. Yellow, pink, mauve or purple flowers, often mottled, are borne in whorls on the upper parts of the stems.

Found on rocky slopes throughout Namaqualand, the Western and Eastern Cape, Lesotho and Namibia.
Flowers July to November.

There is only one species of *Ballota* in southern Africa and in Namaqualand.

Ballota africana
cat herb, kattekruie

LAMIACEAE

An aromatic perennial herb up to 1m high, with hairy stems square in cross-section. The soft, hairy, stalked leaves, up to 3cm long, are broadly sword-shaped to almost circular; the veins are sunken on the upper surface; the margin is irregularly crenate or toothed. Purple or pink to pale mauve flowers are borne in very dense whorls along the upper stem.

Found along drainage lines on flats or rocky slopes throughout Namaqualand and also widespread in other dry areas of the Northern, Western and Eastern Cape, the Free State and Namibia.
Flowers May to November.

PIETER VAN WYK

There are two species of *Mentha* in southern Africa of which one occurs in Namaqualand.

Mentha longifolia
wild mint, wildekruisement

LAMIACEAE

A straggling, aromatic perennial herb up to 1.5m high, usually hairy. The linear to sword-shaped leaves have an entire or sparsely toothed margin. White flowers are crowded in very dense whorls at the ends of branches.

Found along rivers or seeps in Namaqualand and also elsewhere in southern and northern Africa and Europe.
Flowers November to April.

ANNELISE LE ROUX

ANNELISE LE ROUX

SALVIA
There are about 30 species of *Salvia* in southern Africa of which seven occur in Namaqualand.

Salvia africana-lutea
brown sage, bruinsalie, sandsalie
LAMIACEAE

An aromatic, grey shrub up to 1.5m high. The grey, oblong to elliptic leaves have a toothed margin when borne on the main branches but are entire closer to the ends of the branches. Flowers are borne in clusters at the ends of branches; the calyx is covered with short hairs and dotted with glands; the golden to rusty-brown corolla is up to 5cm long; the upper lip, longer than the flower tube, is hooded.

Extending from coastal sand dunes on the Coastal Plain to rocky hillsides in the Namakwaland Klipkoppe and along the coast to Port Alfred. **Flowers July to October.**

ZELDA WAHL & TED OLIVER

Salvia dentata
bergsalie, blousalie
LAMIACEAE

A much-branched shrub up to 1m high. The aromatic, greyish, stalked leaves, up to 2cm long and 1cm across, are borne in clusters and have minute hairs and an undulating, toothed margin. Blue flowers up to 2.5cm long are borne in groups along the ends of branches; the upper lip is usually shorter than the lower; the calyx is minutely hairy and becomes reddish or purple with age.

Found along drainage lines and on rocky slopes in the Richtersveld, Namakwaland Klipkoppe and on the Coastal Plain as well as in the Giftberg, Bokkeveld, Hantam and Roggeveld Mountains. **Flowers May to January.**

Salvia lanceolata
sandsalie
LAMIACEAE

A much-branched, aromatic shrub up to 1m high. The elliptic to egg-shaped, hairy leaves are up to 3.5cm long; the margin is entire or sometimes toothed. Pale yellow, pale pink, maroon or brown flowers up to 3.5cm long are borne in clusters at the ends of branches; the upper corolla lip is 2cm long.

Found in flat, sandy areas or rocky slopes throughout Namaqualand, extending southwards to the Cape Peninsula and Montagu.
Flowers July to October.

TESSA OLIVER

Salvia verbenaca
verbian sage, salie
LAMIACEAE

A perennial herb up to 50cm high, with hairy, erect branches arising from a woody taproot. The sticky, aromatic leaves, growing mainly in a basal rosette, are large, up to 13cm long and deeply incised, becoming smaller on the flowering stalk. Pink, blue or white flowers up to 1.5cm long are borne in whorls on branched stalks; the upper lip is straight or slightly curved.

Found in dry watercourses and disturbed areas in the Namakwaland Klipkoppe and other winter-rainfall areas in southern Africa, also in Mediterranean countries, elsewhere in Europe, the Canary Islands and Asia. It is probably naturalised in the dry, western half of southern Africa, in Australia and in North America.
Flowers June to September.

LITA COLE

LAMIACEAE **385**

NINA HOBBHAHN

Harveya purpurea subsp. *sulphurea*
geelinkblom
OROBANCHACEAE

A root parasite up to 15cm high, with red stems and leaves. The white to yellow flowers are broadly funnel-shaped; the stigma is white or yellow. When a flower is touched or bruised, the surface becomes bluish-black; the flowers turn black when they dry.

Found in sandy or loamy soil on slopes or flats in the Namakwaland Klipkoppe and the Bokkeveld Mountains, extending to the Western and Eastern Cape. **Flowers September to December.**

RUPERT KOOPMAN

Harveya squamosa
jakkalskos
OROBANCHACEAE

A root parasite up to 15cm high, with hairy, yellow to orange stems and scale-like leaves. The tubular flowers are orange with a yellow throat, hairy on the outside; the stigma is exserted beyond the stamens.

Found in deep, sandy soil, mostly coastal but also in the rest of Namaqualand, extending southwards to the Cape Peninsula. **Flowers September to November.**

Hyobanche glabrata
OROBANCHACEAE

A root parasite up to 10cm high, with scale-like leaves. The red flowers are sparsely hairy; the tube is almost straight and cylindrical; the stigma and stamens are exserted beyond the flower tube.

Found on sandy slopes and flats in Namaqualand, extending southwards to Bredasdorp.
Flowers July to October.

RUPERT KOOPMAN

Hyobanche sanguinea
suikerbos, katnaels, jakkalsblom, rooipoppies, kannie
OROBANCHACEAE

A root parasite up to 10cm high. Bright pink to red flowers, hairy on the outside of the hooded corolla, are clustered close together; the stamens are shorter than the flower tube; the white stigma is exserted slightly beyond the tube. Children like to suck the nectar out of the flower as a delicacy.

Found throughout Namaqualand, extending southwards to the Cape Peninsula, eastwards to Port Elizabeth and also found in Namibia.
Flowers June to November.

ANNELISE ERASMUS

OROBANCHACEAE **387**

ANNELISE LE ROUX

ACANTHOPSIS
There are seven species of *Acanthopsis* in southern
Africa of which five occur in Namaqualand.

Acanthopsis carduifolia
ACANTHACEAE

A spiny shrub up to 25cm high, growing
from a woody base. The basal, narrowly
sword-shaped leaves are deeply lobed;
each lobe is tipped with a long, straw-
coloured spine. White, mauve or blue
flowers are borne on erect stalks among
spine-tipped bracts.

Found on stony slopes in Namaqualand
and the Hantam Mountains.
Flowers July to November.

VIRGINIA WEILER & ANNELISE LE ROUX

Acanthopsis disperma
verneuk-halfmensie
ACANTHACEAE

A stemless, spiny annual herb up to 10cm
high. The basal leaves are hairy and
inversely egg-shaped, tapering into the
stalk; the margin has short spines. Blue
to purple flowers, about 2cm long and
3-lobed, are borne between very spiny
bracts in a densely clustered column; after
flowering, the column of bracts becomes
woody and can remain erect for about a
year after the plant has died.

Found on stony flats and slopes in the
Richtersveld and also elsewhere in dry
areas of the Karoo.
Flowers April to September.

388 ACANTHACEAE

BLEPHARIS
There are about 50 species of *Blepharis* in southern
Africa of which two occur in Namaqualand.

Blepharis furcata
ACANTHACEAE

Very spiny, prostrate or cushion-shaped
shrub up to 30cm high. The pale green
to grey leaves have a strongly recurved
margin with or without a few long spines.
Creamy-white to pale blue or pale purple
flowers, about 2cm long and with brownish
or purple veins, are borne between bracts
with very long, branched spines.

Found on rocky slopes throughout
Namaqualand, extending northwards into
central Namibia.
Flowers May to March.

PIETER VAN WYK

Blepharis macra
ACANTHACEAE

A much-branched, spiny shrub up to 45cm
high. The narrowly elliptic leaves are
about 2cm long, with a few small spines
on the margin. Pinkish-white flowers with
brown hairs are borne between bracts
that have evolved to branched spines that
are twice as long as the flowers.

Found on rocky slopes in the
Richtersveld, Namakwaland Klipkoppe and
also in southern Namibia.
Flowers May to January.

ANNELISE LE ROUX

ACANTHACEAE **389**

ZELDA WAHL

JUSTICIA
There are about 30 species of *Justicia* in southern
Africa of which one occurs in Namaqualand.

Justicia cuneata
bloubos
ACANTHACEAE

A spreading, much-branched shrub up to
60cm high. The leaves are up to 1.4cm long,
elliptic, entire and almost hairless. White
flowers with pink or purple markings are
scattered singly along the branches.

Found on sandy flats in Namaqualand
and in other dry areas of the Western
Cape, Eastern Cape and Namibia.
Flowers July to September.

ZELDA WAHL

MONECHMA
There are about 20 species of *Monechma* in southern
Africa of which five occur in Namaqualand.

Monechma mollissimum
ACANTHACEAE

An erect shrub up to 70cm high, with very
hairy stems. The yellow-green, broadly
elliptic leaves, 1–2cm long, are densely
packed, hairy and entire. The 2-lipped,
mauve flowers have a white throat; the
upper lip is a single lobe and the lower lip
is 3-lobed.

Found in sandy areas, usually near
a dry watercourse, in the Richtersveld,
Namakwaland Klipkoppe and also in
northern Bushmanland and in Namibia.
Flowers July to September.

Monechma spartioides
ACANTHACEAE

A sparsely branched shrub up to 60cm high. The leaves are narrow and pointed. Solitary white or white and violet flowers with violet stripes in the centre are scattered along the stems.

Found on rocky flats and slopes throughout Namaqualand, other Karoo areas and in southern Namibia.
Flowers March to September.

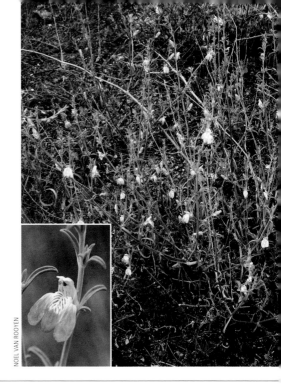

NOEL VAN ROOYEN

Utricularia bisquamata
fonteingesiggie
LENTIBULARIACEAE

A small, carnivorous annual herb up to 5cm high. The stems and leaves are generally reduced or absent and a framework of underwater stolons carry the flower stalks and very small, transparent, bladder-like insect traps. The flowers are white to pale purple to purple, with a raised yellow spot on the lower lip and small, purple-red stripes on the upper petal.

Found in damp, sandy or peaty soil among mosses near streams or wet depressions in the Kamiesberg and throughout southern Africa.
Flowers October to November.

ANNELISE LE ROUX

ACANTHACEAE / LENTIBULARIACEAE **391**

CHASCANUM
There are about 15 species of *Chascanum* in southern
Africa of which three occur in Namaqualand.

Chascanum garipense
VERBENACEAE

A shrub up to 1m high. The oblong
to elliptic leaves are bright green and
stalked; the margin is toothed. Fragrant
white, pale violet or bluish flowers with a
long slender tube and 5 petals are borne
close together on long stalks.

Found on sandy flats, in dry riverbeds
and among boulders on the lower slopes
of hills, in the Richtersveld, extending
eastwards along the Orange River to
Gordonia and also found in southern
Namibia.
Flowers throughout the year after rain.

Chascanum pumilum
VERBENACEAE

A shrub up to 40cm high, with reddish
young branches. The opposite leaves
are elliptic to egg-shaped to oblong;
the reddish margin is toothed. White to
yellow flowers are borne in sparse clusters
at the ends of branches.

Found in dry, sandy riverbeds and
on rocky slopes in the Richtersveld,
Namakwaland Klipkoppe and also in the
dry interior of southern Africa.
Flowers throughout the year after rain.

ANNELISE LE ROUX

WAHLENBERGIA
There are about 180 species of *Wahlenbergia* in
southern Africa of which 22 occur in Namaqualand.

Wahlenbergia acaulis

CAMPANULACEAE   endemic

A annual herb up to 5cm high. The leaves,
covered with long stiff hairs, are scattered
or tufted at the base of the stem; the
margin is toothed. The stalkless, funnel-
shaped flowers are blue with a white
centre and a conspicuous blue dot; the
tube is white with lined blue markings.
     Found in sandy soil on rocky slopes
in the Richtersveld and Namakwaland
Klipkoppe.
Flowers August to September.

MARINDA KOEKEMOER

Wahlenbergia annularis
bluebells, pronkblouklokkie
CAMPANULACEAE

An annual herb up to 20cm high. The
leaves, up to 5cm long, are basal, narrowly
elliptic and hairy; the margin is minutely
toothed. Broadly funnel-shaped, pale
blue, mauve or deep blue flowers are
borne on long, branched, leafless stalks;
the lobes are longer than the tube.
     Found on sandy flats in Namaqualand
and Bushmanland, extending southwards
to Clanwilliam.
Flowers August to November.

ZELDA WAHL

ANNELISE LE ROUX

Wahlenbergia oxyphylla
CAMPANULACEAE

A much-branched, rigid, spiny shrub up to 20cm high. The blue-green, hairy leaves, up to 5mm long, crowded at the lower parts of the branches, are linear, stiff and spiny; the margin is thickened, minutely toothed and rolled inwards. The pale blue to white flowers are about 1cm long, with a tube and 5 lobes.

Found in rock crevices in the Richtersveld and Namakwaland Klipkoppe, the Bokkeveld and Hantam Mountains and the Cederberg, also found in Namibia. **Flowers August to October.**

MARINDA KOEKEMOER

Wahlenbergia polyclada
CAMPANULACEAE   endemic

A prostrate annual herb, completely covered with long, white hairs, the branches up to 15cm long from the centre of the plant. The sword-shaped leaves have a slightly toothed margin. The narrowly funnel-shaped flowers are blue with a dark centre.

Found in deep, wind-blown sand on the Coastal Plain.
Flowers September to October.

Wahlenbergia roelliflora
(previously misidentified as *Wahlenbergia prostrata*)

CAMPANULACEAE   endemic

A prostrate annual herb with branches up to 12cm long from the centre of the plant. The alternate, broadly sword-shaped leaves are folded inwards; the margin is sparsely toothed and hairy. Bell-shaped, pale mauve flowers, with a white centre and 2 purple marks at the base of each petal, are borne on short stalks.

Found on sandy flats in the Namakwaland Klipkoppe.
Flowers in September.

ZELDA WAHL

Wahlenbergia thunbergiana
CAMPANULACEAE

An untidy, brittle shrub up to 80cm high. The linear leaves, up to 1cm long, stand at a 90° angle or are recurved. Star-shaped, pale blue, white or cream-coloured flowers are borne at the ends of branches.

Found on rocky slopes and flats, often within other bushes, in Namaqualand, the Little Karoo and the Great Karoo.
Flowers August to October.

ZELDA WAHL & HEATHER BURGER

CAMPANULACEAE **395**

There are about 15 species of *Monopsis* in southern Africa of which two occur in Namaqualand.

Monopsis debilis
LOBELIACEAE

An annual herb up to 15cm high. The elliptic, basal leaves are up to 5cm long, whereas the leaves on the stems are up to 3cm long; the margin is slightly toothed. Blue-purple flowers are borne in loose clusters at the ends of hairy stalks.

Found in moist areas and along streambanks in the Namakwaland Klipkoppe and also in the Western Cape. **Flowers August to November.**

ANNELISE LE ROUX

CYPHIA
There are about 60 species of *Cyphia* in southern Africa of which six occur in Namaqualand.

Cyphia crenata
bourou
LOBELIACEAE

A deciduous, twining perennial herb with a tuber. The linear to sword-shaped leaves, 2–3cm long, are slightly toothed. Pale purple to pale purple-mauve flowers, up to 1cm long, are normally borne in clusters of 2–5 although they can be solitary; the flower tube is slightly bent; there are 3 smaller and 2 larger, long, pointed lobes, with darker markings in the centre of the lobes.

Found on sandy flats and slopes, twining in other shrubs, in the Namakwaland Klipkoppe, on the Knersvlakte and also extending further south to the Cape Peninsula. **Flowers July to September.**

Cyphia longiflora
bourou

LOBELIACEAE   endemic

A deciduous, twining perennial herb with
a tuber. The slightly fleshy leaves are
linear; the margin is sparsely toothed to
wavy. Flowers, about 1–1.5cm long and
hairy on the back, are borne singly in leaf
axils; the flower tube is slightly curved and
7–8mm long; the petals have crimson-red
to purple markings.
   Found twining in shrubs on rocky
slopes in the Richtersveld and Namakwaland Klipkoppe.
Flowers June to August.

ANNELISE LE ROUX

Cyphia oligotricha
bourou

LOBELIACEAE   endemic

A deciduous, erect perennial herb up to
20cm high, with a tuber. The basal, elliptic
to spathulate leaves are 1–2cm long and
1cm across. The pale mauve flowers are
about 1–1.5cm long; the sharply pointed
lobes are as long as or longer than the
flower tube and only the 3 smaller lobes
have darker markings.
   Found in sand on flats, usually in
drainage lines, on the Knersvlakte.
Flowers July to August.

ANNELISE LE ROUX

LOBELIACEAE **397**

INVOLUCRAL BRACTS FOUND ON THE UNDERSIDES OF FLOWER-HEADS IN GENERA OF THE FAMILY ASTERACEAE

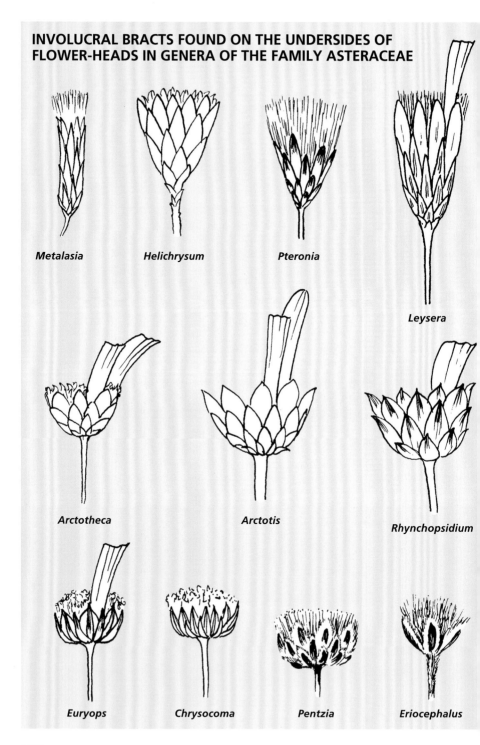

Metalasia

Helichrysum

Pteronia

Leysera

Arctotheca

Arctotis

Rhynchopsidium

Euryops

Chrysocoma

Pentzia

Eriocephalus

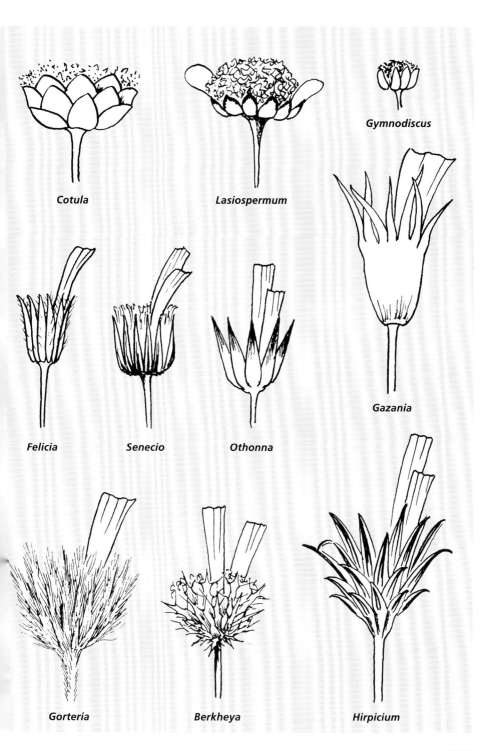

Cotula

Lasiospermum

Gymnodiscus

Felicia

Senecio

Othonna

Gazania

Gorteria

Berkheya

Hirpicium

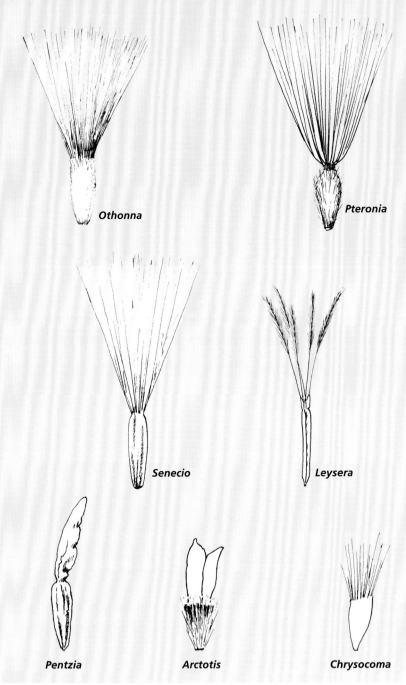

Othonna

Pteronia

Senecio

Leysera

Pentzia

Arctotis

Chrysocoma

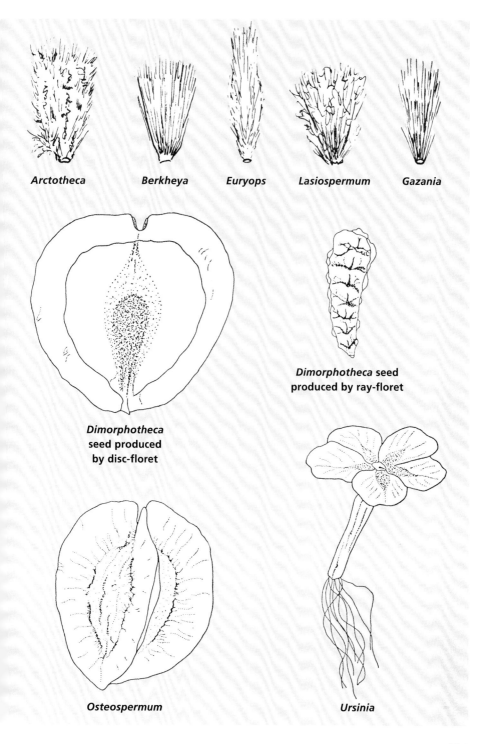

Arctotheca

Berkheya

Euryops

Lasiospermum

Gazania

Dimorphotheca seed produced by ray-floret

Dimorphotheca seed produced by disc-floret

Osteospermum

Ursinia

NORBERT JÜRGENS

DICOMA
There are about 15 species of *Dicoma* in southern Africa of which one occurs in Namaqualand.

Dicoma capensis
koorsbossie

ASTERACEAE

A prostrate perennial herb with stems up to 50cm long. The grey-green, narrowly elliptic to narrowly inversely sword-shaped leaves are covered with greyish, cobwebby hairs above and paler, woolly hairs below; the margin is finely undulate. Pale mauve flowerheads are borne at the ends of branches; the involucral bracts are spiky. The plant is used to treat fever, hypertension and diarrhoea.

Found in sandy soil, usually in dry riverbeds, in the Richtersveld and Namakwaland Klipkoppe and other dry areas of South Africa.
Flowers May to October.

ZELDA WAHL

ARCTOTHECA
There are five species of *Arctotheca* in southern Africa of which two occur in Namaqualand.

Arctotheca calendula
Cape dandelion, soetgousblom

ASTERACEAE

An annual herb up to 15cm high. The leaves are variously lobed, rough above and white-woolly below. Flowerheads up to 6cm in diameter are borne singly on leafless, white-woolly stalks; the ray-florets are pale yellow above and occasionally greenish-grey below; the disc-florets are darker yellow; the involucral bracts are free and in 3 rows, the innermost bracts are the largest and the small outer ones have recurved tips.

Found in sandy areas, often on disturbed sites, in Namaqualand, extending into the Western and Eastern Cape.
Flowers July to September.

402   ASTERACEAE

Arctotheca populifolia
sea pumpkin, seepampoen, strandgousblom
ASTERACEAE

A creeping, mat-forming annual or perennial herb up to 20cm high. The slightly fleshy, white-felted leaves are stalked and heart-shaped, but sometimes lobed at the base; the margin is entire to shallowly dentate. Yellow flowerheads, up to 3cm in diameter and with ray- as well as disc-florets, are borne singly on stalks; the involucral bracts are woolly, with a papery margin. The plants on the beaches along the Namaqualand coast are much smaller and often annual, compared to the more robust perennial plants along the coast elsewhere in South Africa.

Found in shifting sand at the foot of foreshore dunes all along the coast of South Africa.
Flowers July to October.

ANNELISE LE ROUX

Arctotis auriculata
ASTERACEAE

A perennial herb up to 80cm high. The hairy, lobed, blue-green leaves are linear, 3–5cm long. Yellow to orange flowerheads, 4–5cm in diameter and with ray- as well as disc-florets, are borne singly on long stalks; the outer involucral bracts are densely felted, with reflexed tips.

Found in sandy or rocky soil in Namaqualand and Namibia.
Flowers August to October.

ZELDA WAHL & ROBERT MCKENZIE

ASTERACEAE **403**

Arctotis campanulata
(previously misidentified as *Arctotis undulata*)
dubbelgousblom
ASTERACEAE

A deciduous, tufted perennial herb up to 30cm high. The large, inversely egg-shaped, silvery-green, basal leaves are borne on a long stalk and are lightly felted on both surfaces; the margin is slightly lobed and undulating. Deep orange flowerheads are borne on long stalks; the ray-florets are in two series, the inner series with a conspicuous dark brown blotch at the base; the inner involucral bracts are smooth, often dark red-purple and the outer bracts are linear, densely felted and with a reflexed tip.

Found in sandy soil on hills or mountain slopes in the Namakwaland Klipkoppe and on the Bokkeveld plateau, extending southwards to Gouda and Ceres. **Flowers August to September.**

ANNELISE LE ROUX

Arctotis canescens
ASTERACEAE   endemic

A deciduous, tufted perennial herb up to 25cm high. The narrow, blue-green, deeply lobed and undulating leaves are borne on a stalk up to 10cm long and are densely felted on both sides. Deep orange flowerheads are borne on long stalks; the ray-florets are in two series, the inner series with a conspicuous dark brown blotch at the base; the inner straw-coloured involucral bracts are almost smooth and the outer ones are linear, woolly and reflexed, with tips up to 9mm long.

Found in sandy or gravelly soil in the Namakwaland Klipkoppe. **Flowers June to October.**

ANNELISE LE ROUX

Arctotis decurrens
(formerly *Arctotis scullyi*)
wit-sôe, gousblom
ASTERACEAE   endemic

A perennial herb up to 50cm high. The leaves are sword-shaped, up to 13cm long and densely glandular, with soft or rough hairs on both surfaces; the margin is coarsely dentate. White flowerheads about 7.5cm in diameter are borne singly on short leafless stalks; the ray-florets have a black base above and are pink on the underside, while the disc-florets are purple-brown to black; the inner involucral bracts are large and smooth, but the outer bracts have a long, woolly tip and hairy margin.
    Found in sandy soil on the Coastal Plain. **Flowers August to September.**

ZELDA WAHL & ROBERT MCKENZIE

Arctotis fastuosa
bittergousblom
ASTERACEAE   endemic

An annual herb up to 50cm high, branching from the base. The deeply lobed leaves are up to 15cm long, glandular, roughly hairy or loosely felted on both surfaces. Flowerheads up to 6cm in diameter are borne singly on hollow, leafy stalks; the deep orange ray-florets are in two series, the inner series with a conspicuous dark brown blotch at the base; the disc-florets are black when closed and yellow when open; the involucral bracts are in 5 or 6 rows, the outer bracts glandular, softly hairy and with a reflexed, broadly linear tip, while the innermost involucral bracts are the largest, with conspicuous transparent tips.
    Found on sandy flats in Namaqualand. **Flowers July to October.**

ZELDA WAHL

Arctotis revoluta
(formerly *Arctotis laevis*)
krulblaargousblom
ASTERACEAE

A shrub up to 1.5m high. The glandular leaves are twice divided, grey below, smooth or rough above and are aromatic when crushed; the margin is rolled backwards. Flowerheads up to 6cm in diameter are borne singly on long, white-woolly stalks; the ray-florets are yellow above and red on the reverse and the disc-florets are black or yellow; the outer involucral bracts are glandular and narrowly sword-shaped, with reflexed tips.

Found on rocky slopes in Namaqualand, extending southwards to Worcester.
Flowers August to October.

ZELDA WAHL

Arctotis subacaulis
ASTERACEAE

An erect annual herb up to 50cm high. The stalked, deeply lobed leaves are glandular and roughly hairy on both surfaces. Flowerheads are orange or yellow; the ray-florets are in two series, the inner series with a conspicuous dark brown blotch at the base; the outer involucral bracts are glandular and softly hairy, with a short tip that is slightly reflexed.

Found in sandy soil on flats and rocky slopes and in seasonal watercourses in the Namakwaland Klipkoppe and on the Knersvlakte, also in the Roggeveld Mountains and northern Little Karoo.
Flowers August to October.

ROBERT MCKENZIE

There are about 90 species of *Berkheya* in southern
Africa of which seven occur in Namaqualand.

Berkheya fruticosa
vaalperdebos
ASTERACEAE

An erect shrub up to 1.5m high. Young
branches are white-felted. The alternate,
egg-shaped leaves are up to 4cm long and
2.5cm across; the underside is densely white-
woolly and the upper surface is hairless; the
margin is toothed and has spines of up to
4mm long. Yellow flowerheads 3.5–5cm in
diameter are borne singly or in small groups
at the ends of branches; the ray- and disc-
florets are yellow; the involucral bracts are
free, narrow, white-woolly and spiny.

Found in rocky and sandy soil on
hillsides and coastal dunes in Namaqualand,
extending southwards to the Citrusdal area
and also to the Bokkeveld plateau.
Flowers July to October.

PIETER VAN WYK

Berkheya spinosissima
ASTERACEAE

A branched, erect shrub up to 1.7m
high. The stems are glandular and
cobwebbed when young and are usually
tinged purplish. The leaves are opposite,
egg-shaped to oblong, up to 9cm long,
shallowly lobed and almost hairless; the
lobes on the leaf have pale spines up
to 1cm long on the margin and midrib.
Yellow flowerheads about 3.5cm in
diameter are usually borne singly at the
ends of branches; the ray- and disc-florets
are yellow; the involucral bracts of the
outer row are similar to the leaves.

Found in sandy soil on rocky slopes in
Namaqualand, extending into Namibia.
Flowers August to October.

ZELDA WAHL

ANDRE ROMPEL & HEATHER BURGER

There are two species of *Didelta* in Namaqualand and southern Africa.

Didelta carnosa
perdeblom
ASTERACEAE

An annual or biennial herb up to 50cm high. The alternate, slightly fleshy leaves are linear-oblong or inversely sword-shaped, up to 10cm long, entire and hairless. Yellow flowerheads, 4–7cm in diameter and with ray- as well as disc-florets, are borne singly on leafless stalks at the ends of branches; the broadly sword-shaped outer involucral bracts and the smaller, but longer, inner involucral bracts are leaf-like. A highly palatable plant to stock in winter and in summer when the leaves are dry.

Found in sandy areas in Namaqualand, extending southwards to Darling and also found in Namibia.
Flowers July to October.

ZELDA WAHL

Didelta spinosa
perdebos, t'arda
ASTERACEAE

An erect, branched shrub up to 1.5m high and cobwebby on young parts. The bright green, opposite leaves are egg-shaped to elliptic, up to 7cm long, 6cm across and hairless; the margin may have a few spines. The bright yellow to orange flowerheads are 4–7cm in diameter; the ray- and disc-florets are yellow; the outer row of involucral bracts are similar to the leaves, while the bracts of the inner row are smaller. A highly palatable plant to stock, especially the dry leaves in the summer months.

Found in rocky areas in Namaqualand, extending southwards to Piketberg and also found on the Nieuwoudtville plateau.
Flowers July to September.

There are 17 species of *Gazania* in southern Africa of which nine occur in Namaqualand.

Gazania heterochaeta
botterblom
ASTERACEAE

A deciduous perennial herb up to 10cm high. The leaves, in a basal tuft, are inversely egg-shaped, entire or lobed; the upper surface is hairy, whereas the lower surface is white to grey with dense, short, felted hairs. Flowerheads are borne singly on stalks; the ray-florets are yellow to orange with a black mark at the base; the fused part of the involucral bracts has bristle hairs; the free ends of the inner involucral bracts are triangular and smooth and the free part of the outer bracts is hairy.

Found in sandy, stony soil in Namaqualand, extending into the Great Karoo. **Flowers July to September.**

ANNELISE LE ROUX

Gazania jurineifolia
ASTERACEAE

A deciduous perennial herb up to 10cm high. The leaves, in a basal tuft, are deeply segmented almost to the midrib into spine-tipped subtriangular lobes; the leaf surface is white-felted below and sometimes also cobwebby above. Flowerheads are borne singly on short stalks; the ray-florets are white with grey or brown stripes below; the fused involucre is bell-shaped and flat at the base, hairless or with a mealy or sometimes cobwebby appearance; the free ends of the involucral bracts are egg-shaped to triangular and sharply pointed.

Found on sandy or gravelly plains or in shallow soil over rocks in the Richtersveld and Namakwaland Klipkoppe, extending eastwards to Kimberley, southwards to Willowmore and northwards into Namibia. **Flowers July to March.**

PIETER VAN WYK

Gazania krebsiana
botterblom
ASTERACEAE

A deciduous perennial herb up to 20cm high. The leaves, in a basal rosette, are inversely egg-shaped, entire or lobed; the upper surface is hairless to sparsely covered with bristles and sometimes also cobwebby, whereas the lower surface is white-felted. Flowerheads are borne singly on short stalks; the ray-florets are yellow to orange, rarely red, and often have greenish, brownish or greyish bands at the base; the fused part of the involucral bracts is hairless to hairy or sometimes has a mealy appearance.

Found on flats or lower, rocky slopes in the Kamiesberg and on the Knersvlakte, also throughout southern Africa, extending to Tanzania.
Flowers August to September.

ANNELISE LE ROUX

Gazania leiopoda
hondemelkbos, botterblom
ASTERACEAE   endemic

A deciduous perennial herb up to 8cm high, with no or short stems. The leaves, in a basal rosette, are deeply segmented to the midrib; the upper leaf surface has bristle hairs, whereas the lower surface is covered with short, soft, felted, white hairs. Flowerheads are borne singly on short stalks; the deep orange ray-florets have brownish patches with white, often heart-shaped, markings at the base; the cup-shaped, fused part of the involucral bracts is covered with soft hairs and the free tips are triangular.

Found in sandy soil on flats or slopes in the Namakwaland Klipkoppe and also in the rest of the Western Cape.
Flowers August to October.

ANNELISE LE ROUX

Gazania lichtensteinii
botterblom, kougoed
ASTERACEAE

An annual herb up to 25cm high, branching from the base. The inversely sword-shaped leaves are sometimes lobed, usually up to 4cm long, with marginal spines and cobweb-like hairs on both surfaces. Flowerheads 2–4cm in diameter are borne singly on long hairless stalks; the ray-florets are in a single row, yellow or orange, with black marks at the base and greenish stripes on the underside; the smooth involucral bracts are fused for more than two-thirds of their length and the base is flat.

Found on sandy or rocky flats in Namaqualand, Bushmanland, other Karoo areas and in Namibia.
Flowers July to October.

ANNELISE LE ROUX

Gazania othonnites
ASTERACEAE

A mat-forming perennial herb up to 20cm high. The hairless, slightly fleshy, blue-green leaves are sometimes slightly lobed; the margin has spaced bristle hairs. Yellow to orange flowerheads are borne singly on stalks longer than the leaves; the involucral bracts are fused into a bell shape, smooth and collared below.

Found on stony slopes, often between white quartz pebbles, in Namaqualand, extending southwards to Laingsburg.
Flowers August to September.

CAROLIN MAYER

PIETER VAN WYK

Gazania splendidissima
strandbotterblom

ASTERACEAE   endemic

A tufted perennial herb up to 20cm high, with very short, woody branches. The slightly fleshy, deeply lobed leaves have a mealy appearance and are white-felted below, with bristles restricted mainly to the midrib and margin. Flowerheads up to 6cm in diameter are borne on stalks up to 10cm long; the yellow to orange ray-florets have black marks at the base; the grey, velvety involucral bracts are fused for three-quarters of their length and are covered with long bristles that are black at the base.

Found in white sand close to the sea on the Coastal Plain.
Flowers August to October.

ZELDA WAHL

Gazania tenuifolia

ASTERACEAE

A small, prostrate annual herb. The leaves, in a basal rosette, are narrowly linear, deeply lobed into narrow segments, entire and white-hairy below. Flowerheads are 1–2cm in diameter; the yellow to pale orange ray-florets have a brown to black base; the cylindrical fused part of the involucral bracts is hairy between the ribs and has a prominent ridge that surrounds the base.

Found in sandy areas in Namaqualand, extending to the Tankwa Karoo and also found in Namibia.
Flowers July to September.

GORTERIA

There are three species of *Gorteria* in southern Africa of which two occur in Namaqualand.

Gorteria diffusa subsp. *calendulacea*

Kamiesberg beetle daisy

ASTERACEAE   endemic

A prostrate annual herb up to 5cm high, with rough stems on the ground up to 25cm long. The leaves are narrow, inversely sword-shaped, 2–5cm long and hairy on both surfaces. Flowerheads 2–5cm in diameter are borne singly at the ends of branches; the deep orange ray-florets are broad and round and in 2 rows, the inner ones with conspicuous black marks with green and white spots; the free involucral bracts are in many rows; the outer bracts are very narrow and have silky hairs.

Found in sandy soil in the Kamiesberg. **Flowers July to October.**

ZELDA WAHL

Gorteria diffusa subsp. *diffusa*

beetle daisy

ASTERACEAE

A prostrate annual herb up to 10cm high, with rough stems. The narrow, inversely sword-shaped leaves are usually unlobed, up to 5cm long and hairy on both surfaces. Flowerheads 2–3.5cm in diameter are borne singly at the ends of branches; the ray-florets are yellow or orange above, purplish below and have a conspicuous black mark with green and white spots, on 2, usually 3, or sometimes on all of the ray-florets; the free involucral bracts are in many rows; the outer bracts are very narrow and have silky hairs.

Found in sandy soil in Namaqualand, extending to the Olifants River and Breede River Valley, also found in southern Namibia. **Flowers August to October.**

HEATHER BURGER

ASTERACEAE **413**

HIRPICIUM
There are nine species of *Hirpicium* in southern Africa of which two occur in Namaqualand.

Hirpicium alienatum
haarbos
ASTERACEAE

A densely branched shrub up to 50cm high. The rigid, linear leaves are spine-tipped, rough above and white-woolly below; the margin, with long bristles, is rolled back. Flowerheads about 3cm in diameter are borne singly at the ends of leafy branches; the ray-florets are yellow above and brown below; the white silky-haired involucral bracts are partly fused. A highly palatable plant to stock.

Found in rocky areas in Namaqualand and also in other dry areas of the winter-rainfall region in the Northern, Western and Eastern Cape.
Flowers July to October.

ZELDA WAHL & ROBERT MCKENZIE

Hirpicium echinus
ASTERACEAE

A prostrate annual or perennial herb up to 15cm high, branching from the base. The rough leaves are up to 3cm long and lobed into narrow segments, the lobes and tips ending in white spines. Flowerheads 2–3.5cm in diameter are borne singly on short leafless stalks; the ray-florets are white to yellow, with a darker stripe on the undersurface; the disc-florets are yellow; the involucral bracts are in many rows, fused at the base, the outer rows spine-tipped.

Found in sandy soil in Namaqualand and also in Bushmanland, the Kalahari and Namibia.
Flowers July to September.

ANNELISE LE ROUX

There are two species of *Hoplophyllum* in southern Africa of which one occurs in Namaqualand.

Hoplophyllum spinosum
volstruisdoring
ASTERACEAE

A rigid, much-branched, spiny shrub up to 1.5m high, with finely grooved green branches. The spine-like, hard, linear leaves spread at almost 90° from the branch. Yellow flowerheads, with only disc-florets, are crowded at the ends of branches. Fruits have long bristles. A palatable plant to stock.

Found on sandy flats and in riverbeds or on stony slopes in the Richtersveld, on the Coastal Plain and Knersvlakte and in the Hantam and southern Karoo.
Flowers August to December.

ANNELISE LE ROUX

CINERARIA
There are 30 species of *Cineraria* in southern Africa of which three occur in Namaqualand.

Cineraria canescens
wild cineraria, wilde-cineraria
ASTERACEAE

A much-branched shrub up to 50cm high. The very hairy stems are woody towards the base, often with a jointed appearance. The greyish, hairy leaves are kidney-shaped and lobed into small segments at the base, but with a much larger, rounded terminal lobe; the margin is toothed, often crisped. Flowerheads have yellow ray- and disc-florets and are borne on a hairy stalk; the 8–12 cobwebby involucral bracts are pointed and have a membranous edge.

Usually found in rock crevices or between large rocks in Namaqualand, extending to Ceres and Graaff-Reinet, also found in Namibia.
Flowers July to October.

ANNELISE LE ROUX

SENECIO
There are about 310 species of *Senecio* in southern
Africa of which 34 occur in Namaqualand.

Senecio arenarius
pershongerblom
ASTERACEAE

An annual herb up to about 30cm high.
The leaves are up to 7cm long and lobed
into segments; glandular hairs present;
the margin is sometimes rolled backwards.
Flowerheads are 2.5–3cm in diameter; the
ray-florets are magenta; the disc-florets
are yellow; the hairy involucral bracts are
free and in a single row. An unpalatable
plant to stock that is grazed only when
the animals are really hungry – hence
'hongerblom', meaning hunger flower.
Found in sandy areas in Namaqualand,
extending southwards to Agulhas, also
found in Namibia.
Flowers July to September.

ZELDA WAHL

Senecio bulbinifolius
kraaltjies
ASTERACEAE

A dwarf succulent with short, erect stems
up to 10cm high and spreading by means
of arching, leafless or sparsely leafy
runners. The cylindrical, fleshy leaves are
yellow-green with darker green dorsal
stripes. Fragrant yellow flowerheads are
borne on stalks up to 15cm long; the
involucral bracts form a cylinder.
Found on rocky flats or slopes,
often among white quartz pebbles,
in Namaqualand, Bushmanland and
Namibia.
Flowers August to September.

ANNELISE LE ROUX

416   ASTERACEAE

Senecio cardaminifolius
geelhongerblom
ASTERACEAE

An annual herb up to 45cm high. The almost hairless leaves are up to 6cm long and variously lobed, with the leaf bases clasping the stem. Flowerheads about 2cm in diameter are borne in loose clusters; the ray- and disc-florets are yellow; the involucral bracts are free and in a single row.

Found in sandy soil, often in disturbed areas, in Namaqualand, extending through the western Karoo to Worcester and Laingsburg, also found in southern Namibia. **Flowers July to September.**

ZELDA WAHL

Senecio cicatricosus
ASTERACEAE   endemic

A dwarf succulent with a short, knobbly stem up to 10cm high and tuberous roots. The cylindrical, fleshy, blue-green leaves are crowded together and are often curved. White to mauve flowerheads without ray-florets are borne on a smooth stalk scarcely longer than the leaves; the involucral bracts form a cylinder.

Found in rock crevices on dolomite hills on the Knersvlakte. **Flowers March to September.**

ANNELISE LE ROUX

ASTERACEAE **417**

Senecio cinerascens
vieroulap, handjiebos
ASTERACEAE

A perennial shrub up to 1.5m high, with white-woolly stems. The greyish leaves are up to 10cm long, deeply lobed and white-woolly; the margin is rolled back. Yellow flowerheads about 2cm in diameter are borne in groups of 5–10 at the ends of branches; there are only a few ray-florets but many disc-florets; the involucral bracts are almost free and in a single row.

Found on rocky outcrops in the Richtersveld and Namakwaland Klipkoppe, extending to the Cederberg, Hantam and Roggeveld Mountains, also found in southern Namibia.
Flowers July to September.

HEATHER BURGER

Senecio longiflorus
sjambokbossie
ASTERACEAE

A many-stemmed, succulent shrub up to 1m high. The grey-green, erect or sprawling stems are cylindrical, sometimes furrowed, and leafless when flowering. The fleshy leaves are oblong-elliptic. Flowerheads with only white or yellow disc-florets are borne at the ends of branches; the 5 or 6 involucral bracts are fused into a cylinder.

Found on sandy flats and rocky slopes in the Richtersveld, Bushmanland, Karoo and other dry areas of southern Africa.
Flowers August to December.

ANNELISE LE ROUX

Senecio radicans
bobbejaantoontjies, vingertjies
ASTERACEAE

A creeping perennial succulent rooting along the stems. The fleshy, ovoid leaves have a sharp tip, a thick, transparent dorsal stripe and several thin longitudinal stripes. Fragrant white or mauve flowerheads with only disc-florets are usually borne singly on stalks with a few small bracts; the slightly fleshy involucral bracts are arranged in a cylinder around the disc-florets.

Found in rocky areas throughout western, central and southern South Africa and in southern Namibia. **Flowers April to September.**

IVAN VAN NIEKERK

Senecio sisymbriifolius
ASTERACEAE

An erect to lax annual herb up to 40cm high. The thin-textured, hairy leaves are egg-shaped to oblong, lobed, toothed, soft and cool to the touch and sometimes purple below. Yellow flowerheads up to 1.5cm in diameter are borne on a sparsely branched stalk; the involucral bracts are glandular-hairy.

Found in shady, cool places, usually under big boulders, in the Richtersveld and Namakwaland Klipkoppe and in Namibia. **Flowers August to September.**

ZELDA WAHL

EURYOPS
There are about 100 species of *Euryops* in southern Africa of which 10 occur in Namaqualand.

Euryops dregeanus
vaallansrapuis
ASTERACEAE

An erect, much-branched shrub up to 80cm high. The spoon-shaped, velvety-hairy, grey leaves, crowded together in groups on the branches, are up to 4cm long and 1cm across, toothed or lobed at the rounded tips. Yellow flowerheads 3–4cm in diameter are borne singly on yellow-green stalks up to 20cm long; the 8–16 smooth involucral bracts are fused and flat at the base and distinctly 3–5-veined; the wider inner bracts have a membranous margin.

Found on rocky slopes in the Richtersveld, Namakwaland Klipkoppe and on the Knersvlakte, also in the Western Cape and Namibia.
Flowers July to September.

Euryops lateriflorus
soetrapuis
ASTERACEAE

An erect, sticky shrub up to 1.5m high. The grey-green, slightly leathery, elliptic leaves are up to 2cm long, pointed and stalkless. Yellow flowerheads are borne in clusters at the ends of branches; the slightly fleshy involucral bracts are partially fused at the base, in 2 rows, the tips tinged dark blue.

Found at high elevations on mountain slopes in the Namakwaland Klipkoppe, extending to the Bokkeveld, Hantam, Roggeveld and Cold Bokkeveld, also found in southern Namibia.
Flowers August to October.

Euryops multifidus
rapuis
ASTERACEAE

An erect, much-branched shrub up to
1.5m high. The leaves, crowded in groups
mostly at the ends of the branches, are
1–3cm long, deeply divided into needle-
like lobes and hairless. Flowerheads are
yellow and about 1cm in diameter; the 5–9
involucral bracts are fused from the base
to almost halfway along the bracts.
　　Found on rocky slopes in the Richtersveld, Namakwaland Klipkoppe, on the
Knersvlakte and also in the Western Cape.
Flowers July to October.

ZELDA WAHL

Euryops tenuissimus
grootrapuis
ASTERACEAE

An erect shrub up to 2m tall. The needle-
like leaves, usually borne at the ends of
the branches, are 1.5–5cm long. Yellow
to pale orange flowerheads are borne
between the leaves on stalks as long as
the leaves; the 7–24 involucral bracts are
fused from the base to almost halfway
along the bracts.
　　Found at high elevations on mountain
slopes in Namaqualand and other parts of
the winter-rainfall areas in the Northern,
Western and Eastern Cape.
Flowers August to October.

HEATHER BURGER

OTHONNA
There are about 115 species of *Othonna* in southern
Africa of which 30 occur in Namaqualand.

Othonna euphorbioides
ASTERACEAE   endemic

A deciduous succulent up to15cm high, its
thick stems covered with yellow-brown
papery bark. The long-elliptic leaves are
crowded at the branch tips. Tiny, yellow
flowerheads with only disc-florets are
borne on stiff stalks that later become
thorns; there are 5 involucral bracts.

Found in cracks in granite domes in
the Namakwaland Klipkoppe.
Flowers July to November.

HEATHER BURGER

Othonna hallii
ASTERACEAE   endemic

A deciduous succulent up to 10cm high.
The very fleshy, flattened leaves are 2–4cm
long and are tufted at ground level on a
very short, flattened, woolly stem. Yellow
flowerheads with ray- as well as disc-
florets are borne on a smooth stalk up to
10cm long; the 5–8 involucral bracts are
dark-tipped.

Found in white quartz pebble patches
on the Knersvlakte.
Flowers May to August.

ANNELISE LE ROUX

Othonna herrei
ASTERACEAE   endemic

A deciduous succulent up to 15cm high, with fleshy, slightly branched, knobbly stems. The undulating, fleshy leaves, borne on conical knobs consisting of persistent leaf bases, are 3–5cm long and 3cm across; the margin is reddish. Yellow flowerheads with ray- as well as disc-florets are borne on thin, branched stalks; the 6–8 involucral bracts are in a single row and are joined at the base.
    Found on rocky slopes in the Richtersveld.
Flowers May to July.

ZELDA WAHL

Othonna intermedia
bokkool

ASTERACEAE   endemic

A deciduous perennial herb up to 15cm high, with a resinous tuber. The fleshy, blue-green leaves are wedge-shaped and congested at the base of the plant. Yellow flowerheads, about 2cm in diameter and with ray- as well as disc-florets, are borne at the ends of branched stalks; the 8–11 involucral bracts are in a single row and are joined at the base.
    Found between white quartz pebbles on the Knersvlakte.
Flowers June to August.

ANNELISE LE ROUX

ASTERACEAE **423**

NICK HELME

Othonna lepidocaulis
ASTERACEAE   endemic

A deciduous succulent up to 20cm high, covered with hard, glossy, pale yellowish or ivory, scale-like leaf bases (resembling a maize cob). The tufted leaves are inversely sword-shaped, narrowed towards the base, entire or irregularly toothed, and leathery to slightly fleshy. Yellow flowerheads with ray- as well as disc-florets are borne on branched, reddish stalks up to 20cm long; there are about 8 reddish involucral bracts.

Found on stony flats in the Namakwaland Klipkoppe and on the Knersvlakte.
Flowers May to July.

HEATHER BURGER

Othonna macrophylla
bokkool, wildekool
ASTERACEAE   endemic

A deciduous perennial herb up to 60cm high, with a tuber that is felted at ground level. The large, basally congested, blue-green leaves are egg-shaped to elliptic and slightly fleshy to leathery; the margin is sharply toothed. Yellow flowerheads with ray- as well as disc-florets are borne on branched stalks; the 10–15 smooth involucral bracts with reddish tips are joined at the base.

Found on stony slopes in the Richtersveld and Namakwaland Klipkoppe.
Flowers July to September.

Othonna retrorsa
bolletjie mis

ASTERACEAE   endemic

A small, rounded, deciduous dwarf shrub
up to 30cm high, with a long taproot.
The oblong-spathulate, blue-green
leaves, crowded at the ends of very short
branches, are leathery and up to 3cm long;
the slightly thickened margin is finely
toothed. Yellow flowerheads with ray-
as well as disc-florets are borne in loose
clusters on slender stalks about 15cm long;
the 5–8 involucral bracts are in a single
row and are joined at the base, the tips
grey-brown. When the plant has shed its
leaves, the remaining rounded, knobbly,
woody, dark grey stump resembles cow
dung, to which the vernacular name
'bolletjie mis' refers.

Found among flat rocks in the
Namakwaland Klipkoppe.
Flowers August to September.

HEATHER BURGER

Othonna rosea
(previously misidentified as *Othonna cuneata*)

ASTERACEAE   endemic

A deciduous perennial herb up to 40cm
high, with a tuber that is woolly at the
crown. The slightly fleshy, blue-green
leaves are kidney-shaped, shallowly lobed,
faintly veined, about 5cm long and 7cm
across. Mauve to purple flowerheads,
about 2cm in diameter and with ray- as
well as disc-florets, are borne at the ends
of branched stalks; the 8–12 involucral
bracts are in a single row and are joined at
the base.

Usually, but not exclusively, found in
areas with white quartz pebbles in the
Richtersveld and Namakwaland Klipkoppe.
Flowers May to August.

ZELDA WAHL

ASTERACEAE **425**

CRASSOTHONNA
There are 13 species of *Crassothonna* in southern Africa of which nine occur in Namaqualand.

Crassothonna cylindrica
(formerly *Othonna cylindrica*)
ossierapuis
ASTERACEAE

A branched, evergreen shrub up to 1m high. The alternate leaves are fleshy, cylindrical and 3–4cm long. Flowerheads about 2cm in diameter are borne in loose groups above the leaves; the approximately 10 ray-florets are yellow with green veins; the disc-florets are yellow; the 7–9 involucral bracts are in a single row and are joined at the base. An unpalatable plant to stock except for the flowers.

Found in Namaqualand (common on the Coastal Plain) and in Bushmanland, the Western Cape and Namibia.
Flowers July to October.

Crassothonna opima
(formerly *Othonna opima*)
ASTERACEAE

A perennial succulent up to 60cm high. The fleshy leaves are cylindrical, slightly curved, rounded at the tips, 6–10cm long and 1–1.5cm in diameter. Yellow flowerheads, about 1.5cm in diameter and with ray- as well as disc-florets, are borne in loose clusters on long, minutely scaly stalks; the 8–10 involucral bracts are in a single row and are joined at the base.

Found on rocky slopes in the Richtersveld and in southern Namibia.
Flowers August to September.

Crassothonna protecta
(formerly *Othonna protecta*)
ASTERACEAE

A succulent with a tuber, producing small, fleshy, cylindrical or barrel-shaped stems along the soil surface. New rooted plants with crowded leaves are formed along runners. The fleshy leaves are cylindrical, up to 3cm long and 3mm in diameter. Yellow flowerheads are borne on long, fleshy, branched stalks; the 8–10 purplish involucral bracts are in a single row and are slightly fleshy.

Usually, but not exclusively, found among white quartz pebbles on the Coastal Plain and Knersvlakte of Namaqualand, in Bushmanland, the Little Karoo and southern Namibia.
Flowers August to September.

ANNELISE LE ROUX

Crassothonna sedifolia
(formerly *Othonna sedifolia*)
karoorapuis
ASTERACEAE

A rounded, branched shrub up to 50cm high. The fleshy, spherical to short-cylindrical leaves are entire, 4mm in diameter and 5–10mm long. Yellow flowerheads, about 1.5cm in diameter and with ray- as well as disc-florets, are borne on long, slender stalks; the 7–9 involucral bracts are in a single row and are joined at the base. A palatable plant to stock.

Found in sandy soil in Namaqualand, Bushmanland and southern Namibia.
Flowers June to September.

HEATHER BURGER

RIAAN DE VILLIERS

GYMNODISCUS
There are two species of *Gymnodiscus* in southern Africa and both occur in Namaqualand.

Gymnodiscus capillaris
reksening

ASTERACEAE

An annual herb up to 20cm high. The inversely egg-shaped, slightly fleshy leaves, in a basal rosette, are entire or coarsely toothed or lobed. Small, yellow flowerheads, about 6–8mm in diameter and with ray- as well as disc-florets, are borne on sparsely branched, hairless stalks and are arranged in loose clusters. When animals eat this plant their muscles seem to 'stretch' so that they cannot walk for a day or two.

Found on sandy flats in Namaqualand, extending southwards along the west coast and in the Karoo areas of the Western Cape.
Flowers July to September.

ZELDA WAHL

Gymnodiscus linearifolia
reksening

ASTERACEAE

An annual herb up to 20cm high. The fleshy, cylindrical leaves are mostly basal, entire and up to 3cm long. Yellow flowerheads, 4–6mm in diameter and with ray- as well as disc-florets, are borne in clusters at the ends of almost leafless stalks; the 6 involucral bracts are in a single row and are joined at the base. When animals eat this plant their muscles seem to 'stretch' so that they cannot walk for a day or two.

Found in sandy soil in the Namakwaland Klipkoppe and in Bushmanland.
Flowers August to October.

OSTEOSPERMUM
There are about 80 species of *Osteospermum* in southern Africa of which 20 occur in Namaqualand.

Osteospermum moniliferum subsp. *pisiferum*

(formerly *Chrysanthemoides monilifera* subsp. *pisifera*)

sandbietou

ASTERACEAE

An erect, evergreen shrub up to 2m tall, with thinly woolly young branches. The oblong to elliptic leaves are soft, but become leathery with age; the margin is coarsely toothed. The yellow flowerheads are up to 2cm in diameter and have ray- and disc-florets; the inner involucral bracts are broadest above or at the middle. The fleshy fruits become black when ripe.

Found in sandy soil in Namaqualand, extending to the Western and Eastern Cape. **Flowers April to October.**

ANNELISE LE ROUX

Osteospermum grandiflorum

muishondbos

ASTERACEAE

An erect, much-branched shrub up to 1m high. The sticky, alternate, linear to sword-shaped leaves are covered with glandular hairs that emit a foetid smell when the leaf is touched; the margin is toothed. The yellow to orange flowerheads are up to 5cm in diameter and have ray- and disc-florets; the involucral bracts are 3–4 times shorter than the ray-florets.

A highly palatable plant to stock.

Found on rocky slopes in the Richtersveld, Namakwaland Klipkoppe, on the Knersvlakte and also extending southwards to Malmesbury. **Flowers July to October.**

ANNELISE LE ROUX

ASTERACEAE **429**

Osteospermum amplectens
(formerly *Tripteris amplectens*)
dassiegousblom

ASTERACEAE   endemic

An aromatic annual herb up to 50cm high.
The elliptic leaves are up to 10cm long and
sparsely covered with glandular hairs; the
margin is irregularly toothed. The yellow
to pale orange flowerheads are up to
4.5cm in diameter; the fringed involucral
bracts are in 2 rows. Fruits are 3-winged.
    Found in rocky areas in Namaqualand.
Flowers April to November.

ZELDA WAHL

Osteospermum hyoseroides
(formerly *Tripteris hyoseroides*)
dassiegousblom

ASTERACEAE

An aromatic annual herb up to 50cm
high. The elliptic leaves are covered with
short glandular hairs and are up to 10cm
long; the margin is toothed and wavy. The
flowerheads are 3–5cm in diameter, with
orange ray-florets and dark purple disc-
florets; the elliptic involucral bracts have
a transparent margin and are in 2 rows.
Fruits are conspicuously 3-winged.
    Found in stony soils on the
Coastal Plain, in the Richtersveld and
Namakwaland Klipkoppe, also in Bushmanland and the Bokkeveld and Roggeveld Mountains.
Flowers July to September.

ZELDA WAHL

Osteospermum monstrosum
(formerly *Tripteris clandestina*)
trekkertjie
ASTERACEAE

An erect, aromatic annual herb up to
30cm high. The inversely sword-shaped
leaves are up to 8cm long and sparsely
covered with glandular hairs; the margin
is toothed. Pale yellow flowerheads about
1cm in diameter are borne singly or in
small groups at the end of branched
stalks; the elliptic involucral bracts are
in 2 rows and have a broad, transparent
margin. Fruits are conspicuously 3-winged.
    Found in sandy soil in Namaqualand,
in dry areas of the Western Cape and in
southern Namibia.
Flowers July to September.

ZELDA WAHL

Osteospermum microcarpum
ASTERACEAE

An erect annual herb up to 60cm high,
often much branched. The fleshy, elliptic
to inversely sword-shaped leaves are
covered with glandular hairs; the margin
is coarsely toothed and wavy. Yellow
flowerheads with ray- as well as disc-
florets are borne singly or in small groups
at the end of branched stalks. Fruits are
3-winged.
    Found on rocky slopes, and often in or
along dry riverbeds or disturbed sites, in
Namaqualand, extending northwards to
southern Angola.
Flowers throughout the year.

ANNELISE LE ROUX

ASTERACEAE **431**

ZELDA WAHL

Osteospermum oppositifolium
(formerly *Tripteris oppositifolia*)
skaapbos
ASTERACEAE

A round shrub up to 1m high. The slightly
fleshy, hairless, greyish-green leaves are
oblong and 2–4cm long; the margin is
entire. Flowerheads are 4–5cm in diameter
and are borne on short stalks; the ray-
florets are pale yellow near the coast, but
yellowish-orange further inland; the disc-
florets are purple-brown; the involucral
bracts have a narrow, transparent
margin and are in a single row. Fruits are
conspicuously 3-winged. A palatable plant
to stock, especially when in flower.
    Found in sandy or rocky areas in
Namaqualand, extending southwards
to Clanwilliam and also found in
southern Namibia.
Flowers July to October.

ADRIAN FORTUIN

Osteospermum sinuatum
(formerly *Tripteris sinuata*)
kleinskaapbos
ASTERACEAE

A spreading shrub up to 50cm high. The
slightly fleshy, greyish-green leaves are
about 4cm long and oblong to inversely
egg-shaped; the margin is variously
toothed. Yellow flowerheads, up to
3.5cm in diameter and with ray- as well
as disc-florets, are borne on short stalks;
the involucral bracts have a transparent
margin and are in a single row. Fruits
are conspicuously 3-winged. A highly
palatable plant to stock and therefore
grows most successfully in areas where
there is no grazing, such as roadsides.
    Found in sandy or rocky soil in
Namaqualand, in other dry areas of the
Northern, Western and Eastern Cape and
in southern Namibia.
Flowers July to October.

432   ASTERACEAE

DIMORPHOTHECA
There are about 20 species of *Dimorphotheca* in southern Africa of which seven occur in Namaqualand.

Dimorphotheca cuneata
bosmagriet
ASTERACEAE

A round shrub up to 60cm high, with the young parts densely gland-dotted and sticky. The inversely sword-shaped leaves are about 2cm long; the margin is toothed. Flowerheads are borne singly on long stalks at the ends of branches; the ray-florets are shiny white or, rarely, in the Kamiesberg, yellow to orange, and they are darker on the underside; the involucral bracts are 6–9mm long.

Found on rocky slopes or sandy flats in the Kamiesberg, the Roggeveld Escarpment and the Tankwa Karoo, extending to Uitenhage.
Flowers May to September.

RIAAN DE VILLIERS

Dimorphotheca pluvialis
Cape rain daisy, wit-sôe
ASTERACEAE

An erect to sprawling annual herb up to 25cm high, often branching from the base. The oblong leaves, with or without glandular hairs, can be up to 4cm long; the margin is bluntly toothed. Flowerheads, 4–7cm in diameter and with ray- as well as disc-florets, are borne singly on stalks up to 15cm long; the ray-florets are white above and violet below; the free and narrow involucral bracts are in a single row.

Found in sandy areas, often in disturbed sites, in Namaqualand, extending into the Western Cape.
Flowers July to October.

ROBERT MCKENZIE

ASTERACEAE **433**

Dimorphotheca pinnata
(formerly *Osteospermum pinnatum*)
ASTERACEAE

A sticky, prostrate annual herb up to 10cm high. The deeply and finely lobed leaves are up to 3cm long. Flowerheads up to 5cm in diameter are borne singly at the ends of short stalks; the numerous ray-florets are white to orange above (sometimes differing even in the same population), dark purple-brown at the base and dark blue below; the free, narrow, glandular involucral bracts are in a single row. Fruits have thorny edges.

Found in sandy areas in Namaqualand, Bushmanland, the Upper Karoo, Roggeveld, Tankwa Karoo, Little Karoo and southern Namibia.
Flowers July to October.

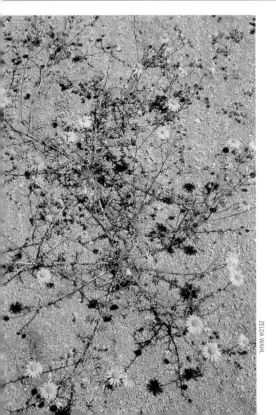

Dimorphotheca polyptera
t'naai-opslag
ASTERACEAE

A glandular annual herb up to 25cm high, branching from the base. The sticky leaves are once- or twice-lobed into very narrow segments and are up to 4.5cm long. Yellow flowerheads, up to 2cm in diameter and with ray- as well as disc-florets, are borne in loose groups at the ends of the stalks; the white-fringed involucral bracts are in a single row.

Found in sandy areas in Namaqualand, extending eastwards into Bushmanland and also found in southern Namibia.
Flowers March to November.

Dimorphotheca sinuata
Namaqualand daisy, jakkalsblom

ASTERACEAE

A loosely branched annual herb up
to 30cm high. The leaves are usually
shallowly lobed, each lobe with a blunt
tooth or, if unlobed, then with blunt teeth
at the tips. Flowerheads up to 5cm in
diameter are borne singly on stalks; the
ray-florets are orange-yellow and blackish
at the base; the free, narrow involucral
bracts are in a single row. A popular, frost-
resistant garden plant.

Found in sandy areas in Namaqualand,
extending southwards to Saldanha and
also found in southern Namibia.
Flowers May to October.

JEAN SMITH & HEATHER BURGER

Dimorphotheca tragus
jakkalsbos

ASTERACEAE   endemic

A perennial herb up to 35cm high. The
glandular-hairy, blue-green leaves are
oblong and up to 7cm long; the margin is
sparsely toothed. White or mostly yellow
to orange flowerheads, 4–6cm in diameter
and with ray- as well as disc-florets,
are borne singly on leafless stalks; the
involucral bracts are free, in 2 rows and
have long points.

Found in sandy areas or on rocky hills
and mountain slopes in Namaqualand.
Flowers July to October.

VERONICA ESTERHUIZEN

ZELDA WAHL

LEYSERA
There are two species of *Leysera* in southern Africa and both occur in Namaqualand.

Leysera gnaphalodes
teebossie
ASTERACEAE

A shrub up to 50cm high. The hairy leaves are grey to green, very narrow, 2–2.5cm long and 0.5–1mm across. Yellow flowerheads, up to 2cm in diameter and with ray- as well as disc-florets, are borne singly on very slender stalks; the papery involucral bracts are arranged in about 5 rows. Where this species is found in abundance it is an indicator of disturbed veld, usually old cropland.

Found on sandy flats in Namaqualand, in the rest of the winter-rainfall areas of the Western and Eastern Cape, also found in southern Namibia.
Flowers September to October.

PIETER VAN WYK

Leysera tenella
teebossie
ASTERACEAE

A small, spreading annual to biennial herb up to 10cm high. The leaves are needle-like, about 1cm long and thinly woolly; the margin is rolled in. Yellow flowerheads, about 1.5cm in diameter and with ray- as well as disc-florets, are borne on very slender stalks; the papery, transparent involucral bracts have brownish tips and are in 5 rows.

Found on sandy flats in Namaqualand and other dry areas in the Northern, Western and Eastern Cape, also found in southern Namibia.
Flowers August to October.

Oedera genistifolia
kleinperdekaroo
ASTERACEAE

An erect shrub up to 1m high. The alternate, stiff, erect to spreading leaves are oblong, often with recurved tips, and are often gummy above. Yellow flowerheads, 5–8mm in diameter and with ray- as well as disc-florets, are borne on short, hairy stalks in clusters at the ends of branches; the inner involucral bracts are rounded.

Found on rocky mountain slopes in the Namakwaland Klipkoppe, the Bokkeveld and Hantam Mountains and on mountains in the Western and Eastern Cape. **Flowers August to November.**

ANNELISE LE ROUX

Rhynchopsidium pumilum
teebossie
ASTERACEAE

A small, spreading annual herb up to 8cm high. The linear leaves are up to 20cm long and have sticky glands. Yellow flowerheads, about 1.7cm in diameter and with ray- as well as disc-florets, are borne singly on slender stalks; the involucral bracts are in 5 rows, the outer bracts long and green, the inner bracts round, brownish and papery.

Found in sandy areas in Namaqualand, extending southwards through the Bokkeveld Mountains to the Little Karoo. **Flowers August to November.**

NICOLA BERGH

ASTERACEAE **437**

RIAAN DE VILLIERS

METALASIA
There are 52 species of *Metalasia* in southern Africa of which three occur in Namaqualand.

Metalasia densa
blombos
ASTERACEAE

An erect shrub up to 1.5m high, with whitish-woolly young branches. Clusters of small leaves are borne in the axils of long, narrow, twisted leaves. White to pink flowerheads are borne densely clustered at the top of the branches; the involucral bracts are erect or, rarely, spreading; the outer bracts are brown below and the inner bracts longer, white or sometimes brown.

Found on rocky slopes in the Namakwaland Klipkoppe, on the Coastal Plain and also in the rest of South Africa. **Flowers August to September.**

NICOLA BERGH

IFLOGA
There are 12 species of *Ifloga* in southern Africa of which eight occur in Namaqualand.

Ifloga candida
ASTERACEAE   endemic

An erect or decumbent annual herb up to 10cm high, with several stems from the basal crown. The linear leaves are occasionally slightly twisted and are thinly white-hairy above; the margin is rolled in. Flowerheads, 4mm long and with only disc-florets, are borne singly or a few together in leaf axils or at the tip of small side branches; the involucral bracts are smooth and shining white.

Found in sandy or rocky flats or slopes on the Knersvlakte.
Flowers August to September.

Ifloga lerouxiae
(formerly *Trichogyne lerouxiae*)

ASTERACEAE   endemic

A small, sticky, erect annual herb up
to 5cm high, with a few to several
branches from the crown; sand adheres
to the plant. The linear leaves, in tufts,
are up to 1cm long and covered with
thin, cobweb-like hairs. Very small,
bell-shaped flowerheads with only disc-
florets are borne between the leaves;
the involucral bracts are translucent and
pale golden brown.

   Found in deep, red sand on the Coastal
Plain of Namaqualand.
Flowers August to September.

Ifloga paronychioides
ASTERACEAE

A prostrate annual herb with branches up
to 4cm long on the ground. The long, thin
leaves, about 1cm long, are white-hairy
and crowded around the flowerheads;
the margin is rolled in. Whitish to pale
golden brown flowerheads are clustered
together; the involucral bracts are silver-
white to pale golden brown.

   Found on sandy flats or rocky slopes
in the Namakwaland Klipkoppe, on the
Bokkeveld and Roggeveld Escarpments
and in southern Namibia.
Flowers July to September.

ASTERACEAE **439**

MARINDA KOEKEMOER

Ifloga pilulifera
ASTERACEAE

A spreading to erect shrub up to 60cm high. The linear leaves, in stalkless clusters, are white-hairy above and occasionally twisted; the margin is rolled in. White flowerheads with only disc-florets are crowded in dense, roundish clusters at the ends of branches; the involucral bracts are golden brown or straw-coloured.

Found on sandy flats or slopes and beach dunes on the Coastal Plain and in the Namakwaland Klipkoppe, extending to the Olifants River Valley and Ceres. **Flowers August to September.**

ANNELISE LE ROUX

Ifloga polycnemoides
ASTERACEAE

An erect annual herb up to 15cm high, with a few to several branches from the basal crown. The linear leaves, often twisted, are white-hairy above; the margin is rolled in. Flowerheads with only disc-florets are borne in clusters between the leaves; the pointed involucral bracts have a reddish central patch and a whitish margin.

Found on rocky hills or sandy flats in the Namakwaland Klipkoppe, the Bokkeveld, Hantam and Roggeveld Mountains, extending to Worcester. **Flowers August to October.**

ELYTROPAPPUS
There are about 10 species of *Elytropappus* in
southern Africa of which one occurs in Namaqualand.

Elytropappus rhinocerotis
(formerly *Dicerothamnus rhinocerotis*)
renosterbos
ASTERACEAE

An erect, grey or grey-green shrub up to
1.5m high, with short, whip-like branches.
Young stems are shortly and densely
woolly. The small, grey-green to olive-green
leaves are almost reduced to scales and are
tightly pressed to the stem. Inconspicuous
purple flowerheads with only disc-florets
are borne at the ends of lateral branches;
the involucral bracts are straw-coloured.

Found on dry, rocky hillsides or dry,
open, sandy areas in the Richtersveld, on
the Coastal Plain and in the Namakwaland
Klipkoppe and it is widespread in the
Western and Eastern Cape.
Flowers April to September.

RIAAN DE VILLIERS

LASIOPOGON
There are eight species of *Lasiopogon* in southern
Africa of which seven occur in Namaqualand.

Lasiopogon micropoides
ASTERACEAE

A small, prostrate, mat-forming annual
herb with stems up to 10cm long on the
ground. The small, greyish-white leaves
are covered with woolly hairs. Minute
flowerheads with only whitish disc-florets,
occasionally tipped reddish, are congested
at the ends of branches; the involucral
bracts are pale straw-coloured.

Found in sandy soil, often in watercourses, in Namaqualand and Namibia.
Flowers June to October.

ANNELISE LE ROUX

ASTERACEAE **441**

ZELDA WAHL

Helichrysum hebelepis
heuningbos
ASTERACEAE

A perennial shrub up to 1m high, branching from a woody base. The narrow, elliptic leaves are 1–1.8cm long, the underside densely woolly and white; the margin is undulate and slightly rolled back. Small, yellow flowerheads, about 3mm in diameter and with only disc-florets, are borne in dense clusters at the ends of branches; the involucral bracts are numerous, rounded, straw-yellow and are woolly on the back. An unpalatable plant to stock.

Found in mountainous areas in Namaqualand and also in the dry areas of the Western Cape.
Flowers August to September.

ZELDA WAHL

Helichrysum leontonyx
mossienes
ASTERACEAE

A prostrate, grey-woolly annual herb, producing many branches up to 15cm long from the crown. The hairy, oblong-elliptic, grey-green leaves are 1.5–3cm long and 0.5–1cm across. Small flowerheads with only minute disc-florets are borne in small compact clusters; the involucral bracts are in 3 or 4 rows, the outermost short and webbed with wool to the leaves, the inner purple-red and longer than the disc-florets themselves.

Found on sandy flats in Namaqualand and in the Western Cape.
Flowers July to October.

Helichrysum revolutum
ASTERACEAE

An erect or sprawling shrub up to 2m
tall, with greyish-white, woolly young
branches. The linear leaves, 1–4cm long
and borne in tufts with a long leaf
supporting a cluster of shorter leaves,
are slightly hairy above and white-woolly
below; the margin is rolled backwards.
Yellow flowerheads with only disc-florets
are borne in clusters at the ends of
branches; the oblong involucral bracts,
with rounded ends, are in 4 or 5 rows and
pale to deep straw-coloured.
　　Found on rocky or sandy flats and slopes
in Namaqualand, extending to the Cape
Peninsula and the Witteberg Mountains,
also found in southern Namibia.
Flowers July to October.

MARINDA KOEKEMOER

Helichrysum stellatum
ASTERACEAE

An erect to spreading, grey-woolly shrub
up to 45cm high. The grey, woolly, linear-
oblong leaves are up to 3cm long and
5mm across; the margin is flat or crisped.
Broadly bell-shaped, white flowerheads
with only disc-florets are borne singly at
the ends of branches; the outer involucral
bracts are brownish to cream-coloured
and the inner radiating involucral bracts
are whitish to pinkish and papery.
　　Found on sandy flats and slopes
on the Knersvlakte and the Bokkeveld
Mountains, extending to the
Riviersonderend Mountains.
Flowers September to October.

MARINDA KOEKEMOER

Helichrysum tinctum
ASTERACEAE

A prostrate, grey- or sometimes white-woolly annual or perennial herb with congested branches up to 15cm long. The elliptic leaves are up to 1.5cm long and 5mm across, with both surfaces woolly felted; the margin is recurved. Pale to deep straw-coloured to golden brown or crimson flowerheads are borne in small compact clusters; the involucral bracts are in 4 series, the outer ones short and webbed with wool to the surrounding leaves, the inner ones exceeding the flowers.

Found in sandy soil in the Namakwa-land Klipkoppe and Bokkeveld Mountains, extending to Uniondale and Graaff-Reinet. **Flowers September to November.**

ZELDA WAHL

AMELLUS
There are 12 species of *Amellus* in southern Africa of which eight occur in Namaqualand.

Amellus alternifolius
ASTERACEAE

A bristly annual or short-lived perennial herb up to 60cm high. The alternate, linear leaves are up to 4cm long; the margin is entire or 1–3-toothed. Flowerheads with bluish-mauve or white ray-florets and yellow disc-florets are borne singly at the ends of branches; the involucral bracts are roughly hairy and glandular.

Found on sandy flats in the Namakwa-land Klipkoppe and on the Coastal Plain, extending to Lambert's Bay. **Flowers August to November.**

ANNELISE LE ROUX

Amellus microglossus
ASTERACEAE

A hairy, sprawling to prostrate annual herb up to 15cm high. The hairy leaves are elliptic; the margin is entire. Flowerheads with small, white ray-florets and yellow disc-florets are borne singly at the ends of branches; the often red-tipped involucral bracts are covered with long bristles and glands.

Found on sandy flats and in dry watercourses in Namaqualand and the winter-rainfall areas of southern Africa. **Flowers July to October.**

ANNELISE LE ROUX

Amellus nanus
ASTERACEAE

A usually compact, rounded annual herb up to 20cm tall. The alternate, elliptic leaves are covered with hairs pressed closely to the leaf surface; the margin is entire. Flowerheads with mauve to pale blue or white ray-florets and yellow disc-florets are borne singly at the ends of branches; the involucral bracts are covered in bristles and are often purplish-tipped.

Found in sandy soils in the Richtersveld, on the northern Coastal Plain and in southern Namibia. **Flowers August to September.**

HELMUTH KOHRS

There are about 100 species of *Felicia* in southern Africa of which 20 occur in Namaqualand.

Felicia australis
sambreeltjies
ASTERACEAE

An annual herb up to 20cm high and branching from the base. The narrow, linear leaves are up to 1.5cm long; the margin is entire and stiffly hairy. Flowerheads, up to 3cm in diameter and with blue to mauve ray-florets and yellow disc-florets, are borne singly on leafless stalks; the involucral bracts are rough and in 3 rows.

Found on sandy flats in Namaqualand and the western part of the Western Cape. **Flowers July to October.**

ZELDA WAHL

Felicia brevifolia
ASTERACEAE

An erect shrub up to 1.5m high, with a greyish bark. The alternate, hairy, greyish-green leaves are up to 1.5cm long; the margin is entire to variously lobed or indented. Flowerheads, about 3–5cm in diameter and with blue to mauve ray-florets and yellow disc-florets, are borne singly on short leafless stalks; the involucral bracts are in 3 or 4 rows, hairy and glandular. A palatable plant to stock.

Found in sandy or loamy soil on rocky slopes in the Richtersveld, Namakwaland Klipkoppe and in southern Namibia. **Flowers May to September.**

Felicia dubia
sambreeltjies
ASTERACEAE

An erect annual herb up to 25cm high.
The sword-shaped, elliptic or inversely
egg-shaped leaves are hairy; the margin
is weakly toothed. Flowerheads, about
2.5cm in diameter and with blue ray-
florets and yellow disc-florets, are borne
singly on long, hairy stalks; the involucral
bracts are in 3 rows, with the 2 inner rows
of equal length and the outer row shorter.
     Found in sandy or loamy soil on rocky
slopes in Namaqualand, extending into
the Western Cape.
Flowers August to September.

Other very similar-looking annual species
of this genus in Namaqualand, with pale
blue ray-florets and yellow disc-florets, are
F. ergeriana, *F. merxmuelleri*, *F. microsperma*,
F. namaquana, *F. puberula* and *F. tenera*.

ZELDA WAHL

Felicia filifolia
persbergdraaibossie
ASTERACEAE

A much-branched shrub up to about
60cm high. The slightly fleshy, needle-like
leaves are up to 4cm long and are borne
in clusters on the branches. Solitary violet
flowerheads, 3–3.5cm in diameter and
with ray- as well as disc-florets, are borne
on short, leafless stalks at the ends of
branches; the involucral bracts are sword-
shaped and in 3 or 4 rows. A palatable
plant to stock.
     Found on sandy flats or rocky slopes
in the Richtersveld and Namakwaland
Klipkoppe, also widespread from southern
Namibia and Limpopo Province to the
Western Cape.
Flowers August to October.

ZELDA WAHL

CHRYSOCOMA
There are about 20 species of *Chrysocoma* in southern Africa of which 10 occur in Namaqualand.

Chrysocoma ciliata
bitterbos
ASTERACEAE

A much-branched shrub up to 40cm high. The alternate, needle-like leaves are up to 1.5cm long and hairless or with a few short hairs. Yellow flowerheads, about 1cm in diameter and with only disc-florets, are borne singly on short stalks at the ends of branches; the involucral bracts are in 4 rows and have a membranous margin. An unpalatable plant that can cause 'vuursiekte' in sheep, resulting in the animals' wool falling out.

Found in sandy, rocky areas in Namaqualand and also in other dry areas of South Africa.
Flowers August to November.

ZELDA WAHL

NOLLETIA
There are seven species of *Nolletia* in southern Africa of which one occurs in Namaqualand.

Nolletia gariepina
ASTERACEAE

A shrub up to 60cm high, the whole plant densely covered with spreading hairs and occasionally a few small glands. The linear-oblong leaves are sparsely arranged along the branches. Yellow flowerheads with only disc-florets are borne singly at the ends of branches; the involucral bracts are in 3 rows.

Found in rock crevices and on rocky slopes in the Richtersveld and in Namibia.
Flowers August to November.

ANNELISE LE ROUX

PTERONIA
There are about 70 species of *Pteronia* in southern
Africa of which 24 occur in Namaqualand.

Pteronia divaricata
geelknopbos, spalkpenbos
ASTERACEAE

A dense shrub up to about 1.5m high, with
greyish bark. The opposite, broadly elliptic
leaves, with a coarse surface, are up to 1.5cm
long and 1.2cm across. Yellow or whitish
flowerheads, about 4mm in diameter, 1.3cm
long and with only disc-florets, are borne
in clusters at the ends of branches; the
involucral bracts are egg-shaped to linear.
An unpalatable plant to stock.

Found on sandy flats and rocky slopes
in Namaqualand, extending into the
Western Cape and Namibia.
Flowers August to November.

Pteronia undulata differs in having smaller
undulate leaves.

ZELDA WAHL

Pteronia glabrata
knopbos
ASTERACEAE

A spreading perennial shrub up to 60cm
high. The opposite, slightly fleshy, blue-
green leaves are narrowly oblong and up
to 5cm long; the margin is entire. Yellow
flowerheads with only disc-florets are
borne singly at the ends of branches;
the involucral bracts, in 4 or 5 rows, are
roundish with a whitish, papery margin.

Found on sandy flats and rocky slopes
in Namaqualand, extending southwards to
Clanwilliam and into the northern Tankwa
Karoo, also found in southern Namibia.
Flowers July to October.

ZELDA WAHL

Pteronia incana
t'kaibebos
ASTERACEAE

A tangled bush up to 1m high, with dark bark. The opposite, greyish, white-woolly leaves are 1cm long and 2mm across. Yellow flowerheads, about 4mm in diameter, 1cm long and with only disc-florets, are borne singly at the ends of branches; the hairless involucral bracts have a membranous margin and are in 4–10 rows, the inner ones becoming progressively larger.

Found on sandy flats or rocky slopes in Namaqualand and also in other dry areas in the Western and Eastern Cape, extending to Port Elizabeth.
Flowers August to October.

ZELDA WAHL

Pteronia leptospermoides
ASTERACEAE   endemic

A shrub up to 1.5m high, with reddish young branches that become darker brown with age. The opposite, linear leaves are flat or, when very narrow, almost cylindrical. White flowerheads with only disc-florets are borne singly at the ends of branches; the contrasting, exerted styles are purple; the involucral bracts closely overlap each other and grade from green to white.

Found on rocky slopes in the Richters-veld and Namakwaland Klipkoppe.
Flowers August to November.

ANNELISE LE ROUX

Pteronia onobromoides
boegoe
ASTERACEAE

A rounded shrub up to 75cm high.
The fleshy, aromatic, alternate leaves
are elliptic and hairless; the margin
has closely set bristly teeth. Yellow
flowerheads with only disc-florets are
borne singly at the ends of branches;
the outer involucral bracts are egg-
shaped and the inner ones are oblong. A
palatable plant to stock and wildlife.
　　Found on sandy flats on the Coastal
Plain, extending southwards to Saldanha
and also found in southern Namibia.
Flowers November to December.

ANNELISE LE ROUX

Pteronia villosa
ASTERACEAE

A shrub up to 30cm high, with white-
woolly young branches that become
hairless with age. The alternate leaves
are oblong and covered with small
bristle-like hairs; the margin is entire.
Yellow flowerheads with only disc-florets
are borne singly at the ends of branches;
the outer involucral bracts are oblong-
elliptic, with a jagged, membranous
margin, whereas the inner ones are
oblong and rounded.
　　Found in sandy loam and in white
quartz pebbles on undulating plains of the
Knersvlakte and northern Tankwa Karoo.
Flowers August to September.

UTE SCHMIEDEL

COTULA
There are about 45 species of *Cotula* in southern
Africa of which 10 occur in Namaqualand.

Cotula barbata
gansogies, knoppiesopslag
ASTERACEAE

An annual herb up to 15cm high. The very
hairy, deeply divided leaves are crowded in
a basal tuft and are about 4cm long. Yellow
flowerheads, about 1cm in diameter and
with only disc-florets, are borne singly on
long, slender, hairless stalks; the involucral
bracts are almost round and papery, with a
transparent margin.

Found in sandy soil in the
Namakwaland Klipkoppe and also in the
Nieuwoudtville and Clanwilliam areas.
Flowers August to September.

Cotula coronopifolia
gansgras
ASTERACEAE

An erect or trailing, hairless annual
herb up to 20cm high, often rooting at
the nodes. The alternate, slightly fleshy
leaves enfold the stem at the base and
are irregularly toothed and lobed. Yellow
flowerheads with only disc-florets are
borne singly on long, slender stalks.

Found in seasonally damp areas
in Namaqualand and also throughout
southern Africa. This species has been
introduced to Argentina and Australia.
Flowers July to September.

Cotula laxa
knoppiesopslag

ASTERACEAE   endemic

A small, erect annual herb up to 10cm
high, mostly branching from the base.
The alternate leaves are up to 4cm long,
deeply once- or twice-lobed and thinly
hairy. Pale yellow to white flowerheads,
about 4mm in diameter and with only
disc-florets, are borne singly on long,
slender, hairless stalks; the involucral
bracts have a papery margin.

   Found in sandy soil in partly shaded
sites between rocks in the Richtersveld
and Namakwaland Klipkoppe, extending
southwards to Montagu.
Flowers August to September.

Cotula leptalea

ASTERACEAE   endemic

A tufted annual herb up to 10cm high.
The alternate leaves are thinly hairy,
deeply lobed and 1–4cm long. Yellow,
cream-coloured or white flowerheads,
about 8mm in diameter and with only
disc-florets, are borne on minutely hairy
stalks that are much longer than the
leaves; the involucral bracts are in a
single row.

   Found in moist, sandy places on
rocky slopes in the Richtersveld and
Namakwaland Klipkoppe.
Flowers July to October.

ASTERACEAE **453**

ANNELISE LE ROUX

Cotula pedicellata
ASTERACEAE

An annual herb up to 30cm high. The tufted leaves at the base of the plant are softly hairy, deeply once- or twice-lobed into linear segments and sharply pointed. Yellow flowerheads with only disc-florets are borne singly at the ends of long, hairy stalks; the numerous overlapping involucral bracts, in 3 or 4 rows, are covered with long, soft hairs and have membranous tips.

Found in deep sand on the Coastal Plain and Knersvlakte, extending into the Olifants River Valley.
Flowers August to September.

ZELDA WAHL

URSINIA
There are about 40 species of *Ursinia* in southern Africa of which 12 occur in Namaqualand.

Ursinia anthemoides
ASTERACEAE

An erect annual herb up to 30cm high, often with reddish stems. Leaves up to 4cm long, mostly deeply lobed into narrow segments and sometimes twice-lobed at the ends. Flowerheads are borne singly on long, thin stalks; ray-florets salmon to pale yellow, with a dark purple base, purple-brown below; the involucral bracts are arranged in many overlapping rows, outer bracts have a membranous margin, middle ones have triangular tips and inner bracts have roundish tips; 3 or 4 outer rows of bracts often have a dark brown margin.

Found in sandy loam on slopes in Namaqualand, the rest of the Western Cape, Eastern Cape and in southern Namibia.
Flowers July to October.

Ursinia cakilefolia
glansooggousblom

ASTERACEAE   endemic

An annual herb up to 30cm high. The leaves are up to 5cm long, toothed and twice-lobed into very narrow segments with bristles on the tips of the lobes. Orange flowerheads up to 3cm in diameter are borne singly on long, slender stalks; the black-margined involucral bracts are in many rows, the innermost bracts the largest and with conspicuous transparent tips.

Found in sandy soils, often in masses in disturbed areas, in Namaqualand and in the northern parts of the Western Cape. **Flowers July to September.**

RIAAN DE VILLIERS

Ursinia calenduliflora
berggousblom

ASTERACEAE   endemic

An annual herb up to 35cm high. The leaves are up to 5cm long and mostly deeply lobed into narrow segments. Flowerheads 2–5cm in diameter are borne singly on slender stalks; the ray-florets are orange, usually with a purple base, the disc-florets are yellow; the involucral bracts are free and in many rows, the innermost ones the largest, with conspicuous brownish, transparent tips, the outer ones with rounded, membranous tips.

Found in sandy soil, often in disturbed areas, in the Richtersveld and Namakwaland Klipkoppe. **Flowers July to September.**

JEAN SMITH

Ursinia chrysanthemoides
pokkiesbos
ASTERACEAE

An untidy perennial shrub up to 1m high, with young stems herbaceous and white. Occasionally, this shrub can spread on the ground and root at the nodes. The leaves are twice-lobed into very narrow segments. Flowerheads are borne singly on long stalks; the ray-florets are yellow, orange, occasionally white or red and the undersides have a darker, coppery colour; the disc-florets are yellow, sometimes with purple tips; the outer involucral bracts have membranous, triangular, sharply pointed tips and are lacerated; the inner bracts have egg-shaped or rounded, silvery appendages.

Found on sandy flats and rocky slopes in Namaqualand and in the other winter-rainfall areas.
Flowers July to October.

ALTA CHAMBERS

ANNELISE LE ROUX

Ursinia kamiesbergensis
ASTERACEAE   endemic

An annual herb up to 30cm high, branching from the base. The leaves are deeply lobed with a sharp tip on each lobe. Yellow flowerheads with ray- as well as disc-florets are borne singly on sparsely hairy stalks much longer than the leaves; the involucral bracts are arranged in 4–6 rows, with thick, membranous, opaque, cloudy tips; the narrow, sharply pointed outer involucral bracts have a purplish-black line at the base of the membranous tip; the inner involucral bracts have bigger, rounder membranous tips.

Found in sandy soil on the Kamies-berg plateau.
Flowers August to September.

Ursinia nana
kleinmagrietjie
ASTERACEAE

A prostrate annual herb up to about 5cm
high. The hairless leaves are deeply lobed,
1.5–3.5cm long and slightly fleshy. Yellow
flowerheads, up to 2.5cm in diameter and
with ray- as well as disc-florets, are borne
singly on short stalks; the involucral bracts
are broadly triangular, with a dark margin
and blunt tip.

   Found in flat sandy soil in Namaqua-
land and also in other dry areas of South
Africa and in southern Namibia.
Flowers July to September.

Ursinia speciosa
ASTERACEAE

An annual herb up to 40cm high and
branching from the base. The leaves are
deeply twice-lobed into fine segments and
up to 5cm long. Flowerheads about 5cm in
diameter are borne singly on leafless stalks;
the ray-florets are yellow, occasionally
white; the disc-florets are yellow,
occasionally purple; all involucral bracts
have large, transparent, rounded tips.

   Found in deep, sandy soil in
Namaqualand, extending southwards
to Malmesbury and also found in
southern Namibia.
Flowers July to October.

ERIOCEPHALUS
There are 34 species of *Eriocephalus* in southern Africa of which 12 occur in Namaqualand.

Eriocephalus brevifolius
wild rosemary, wilderoosmaryn
ASTERACEAE

A small, much-branched shrub up to 1m high. The short branches bearing the leaves are alternate. The felted, grey-green leaves are mostly small and narrow, covered in silvery hairs. Small, woolly flowerheads are clustered at the ends of branches; there are 3 or 4 white ray-florets and the disc-florets are deep purple; the outer involucral bracts are free, whereas the inner ones are fused. Fruits are covered with long, light brown to white hairs. A palatable plant to stock.

Found on rocky flats and slopes in the Namakwaland Klipkoppe, extending into the Western Cape.
Flowers July to September.

Eriocephalus microphyllus
kapokbos
ASTERACEAE

A much-branched shrub up to 1m high. The short branches bearing the leaves are opposite. The leaves are grey-green, narrow and about 3mm long, giving off a scent of rosemary when crushed. Flowerheads are borne singly, spread out along the upper part of the branches; the white ray-florets are almost hidden in the involucral bracts. Fruits are covered with long, white hairs.

Found on rocky flats and slopes in the Richtersveld and Namakwaland Klipkoppe and in the Bokkeveld, Hantam and Roggeveld Mountains.
Flowers July to August.

Eriocephalus racemosus
strandkapokbos

ASTERACEAE

An erect shrub up to 1.5m high, with long slender branches. The silvery-hairy, linear, slightly fleshy leaves are borne in clusters and are very aromatic. Slightly hanging flowerheads lacking ray-florets are spread out along the branches. Fruits are very fluffy and resemble cotton wool.

Found in deep sand on the Coastal Plain and Knersvlakte, extending southwards and then eastwards along the coast to Humansdorp.
Flowers July to September.

ANNELISE LE ROUX

HYMENOLEPIS
There are seven species of *Hymenolepis* in southern Africa of which one occurs in Namaqualand.

Hymenolepis crithmifolia
(formerly *Hymenolepis parviflora*)
pokbos

ASTERACEAE

An erect shrub up to 1m high. The needle-like, grey-felted leaves are 3–10cm long and deeply divided; the margin is rolled in. Flat, densely packed, yellow flowerheads, with only disc-florets and emitting a strong, pleasant scent, are borne at the ends of branches; the involucral bracts are smooth and dry.

Found on rocky mountain slopes in the Namakwaland Klipkoppe and the Bokkeveld Mountains, extending southwards to Swellendam.
Flowers November to December.

ANNELISE LE ROUX

LASIOSPERMUM
There are four species of *Lasiospermum* in southern Africa of which two occur in Namaqualand.

Lasiospermum brachyglossum
knoppiesopslag
ASTERACEAE

An erect annual herb up to 30cm high. The leaves are twice-lobed into linear segments. Flowerheads 1–1.5cm in diameter are borne singly on slender leafless stalks; the widely spaced ray-florets are white and reddish-purple, short and bent backwards; the disc-florets are yellow and red; the involucral bracts have a membranous margin and are in 3 rows.

Found in sandy areas in Namaqualand and on the inland escarpment, extending to the Little Karoo and also found in southern Namibia.
Flowers July to September.

ZELDA WAHL

Lasiospermum pedunculare
gansgras
ASTERACEAE

A tufted or sprawling perennial herb up to 20cm high, with stems arising from a woody rootstock. The grey leaves are finely divided and covered with silvery-silky hairs. Fragrant yellow flowerheads, button-like and lacking ray-florets, are borne singly on stout, leafy stalks; the involucral bracts are silky, with membranous tips.

Found in clayey loam soils in the Namakwaland Klipkoppe and Bokkeveld Mountains, extending to Middelpos and Ceres.
Flowers August to September.

RIAAN DE VILLIERS

There are three species of *Leucoptera* in southern Africa and all are found in Namaqualand.

Leucoptera oppositifolia

ASTERACEAE   endemic

An erect to spreading, untidy shrub up to 50cm high. The opposite, flattened leaves are thick and leathery; the margin is entire or sometimes 3- or 4-lobed. Flowerheads are borne singly on long stalks; the ray-florets are 3–7mm long, white, becoming pink-reddish and bending backwards when older; the corolla lobes of the yellow disc-florets have a central resin gland; the involucral bracts overlap each other and their tips are rounded.

Found in rocky, quartz patches on the Knersvlakte.
Flowers August to September.

ANNELISE LE ROUX

Leucoptera subcarnosa

ASTERACEAE   endemic

A shrub up to 10cm high, spreading on the ground. The fleshy, cylindrical or flattened leaves are closely set; the margin is entire or sometimes 2–4-toothed or -lobed. Flowerheads are borne singly on long stalks; the 4–10mm-long ray-florets bend backwards and are white, often turning pink or reddish when older; the corolla lobes of the yellow disc-florets do not have resin glands.

Found in rocky, quartz patches in Namaqualand.
Flowers July to August.

MARINDA KOEKEMOER

ASTERACEAE **461**

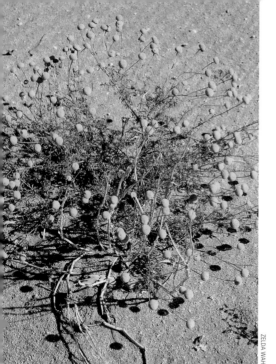

FOVEOLINA
There are four species of *Foveolina* in southern Africa
of which two occur in Namaqualand.

Foveolina dichotoma
stinkkruid, gansogies
ASTERACEAE

An annual herb up to 30cm high,
branching mainly from the base. The
leaves are deeply twice-lobed, sparsely
or densely covered with minute hairs and
up to 2cm long. Almost round, yellow
flowerheads, 5–7mm in diameter and with
only disc-florets, are borne singly on thin,
leafless stalks; the involucral bracts have a
membranous margin and are in 4 rows. An
unpalatable plant to stock.

Found on sandy or sometimes stony
flats and in washes in Namaqualand, other
dry parts of the Northern, Western and
Eastern Cape and in southern Namibia.
Flowers August to September.

Foveolina tenella
ASTERACEAE

An annual herb up to 20cm high. The
thinly hairy, aromatic leaves are deeply
divided into elliptic lobes and are strongly
scented when crushed. Flowerheads
1.5–2.4cm in diameter are borne singly
on long, leafless stalks; the 4–6 white
ray-florets are bent backwards and the
many disc-florets are yellow; the involucral
bracts have a transparent margin and are
arranged in a few overlapping rows.

Found in deep, red sand on the Coastal
Plain and the Knersvlakte, Bokkeveld
plateau and southwards along the west
coast plains to Yzerfontein.
Flowers August to September.

There are 22 species of *Pentzia* in southern Africa of which four occur in Namaqualand.

Pentzia incana
gansogiebos, alsbossie, karoobossie

ASTERACEAE

A rounded, densely branched shrub up to 50cm high, with silvery-felted young branches. The aromatic leaves are deeply divided, 2–6mm long and thinly or densely covered in white hairs. Round, yellow flowerheads, 6mm in diameter, with only disc-florets, are borne singly on stalks at the ends of branches; the involucral bracts have a membranous margin and are in 4 rows. An unpalatable to palatable plant to stock.

Found on rocky flats and slopes in the Richtersveld, Namakwaland Klipkoppe and on the Knersvlakte, extending to the Eastern Cape and Upper Karoo and also found in southern Namibia.
Flowers July to October.

ANNELISE LE ROUX

Pentzia peduncularis
ASTERACEAE   endemic

A spreading shrub up to 50cm high, with densely whitish-felted young branches. The deeply divided, grey-green leaves, crowded at the base of the flowering stalk, are 5–25mm long and sparsely to densely hairy. Round, yellow flowerheads, about 1cm in diameter and with only disc-florets, are borne singly on stout, very hairy stalks up to 30cm long; the narrow, hairy involucral bracts have a narrow, membranous margin.

Found in rocky quartz patches on the Knersvlakte.
Flowers July to August.

VERONICA ESTERHUIZEN

ASTERACEAE **463**

ZELDA WAHL

ONCOSIPHON
There are seven species of *Oncosiphon* in southern
Africa of which four occur in Namaqualand.

Oncosiphon grandiflorus
groot-knoppiesstinkkruid
ASTERACEAE

An erect annual herb up to 40cm high.
The thinly hairy leaves, with a pungent
smell, are finely divided and up to 5cm
long. Yellow flowerheads, up to 1cm in
diameter and with only disc-florets, are
borne singly on short, stout stalks at the
ends of branches; the hairy involucral
bracts have a membranous margin and are
in 4 rows.

Found in sandy soil, often on ploughed
land, in Namaqualand, the Western Cape
and southern Namibia.
Flowers August to October.

HEATHER BURGER

Oncosiphon sabulosus
ASTERACEAE

A rounded annual herb up to 25cm high,
with densely leafy stems. In the south,
this plant can become biennial. The
leathery, often slightly fleshy leaves can be
shallowly to deeply lobed. Round, yellow
flowerheads, 5–7mm in diameter and with
only disc-florets, are borne in clusters at
the ends of branches; the involucral bracts
are almost hairless.

Found in deep sand close to the sea
on the Coastal Plain and along the coast
to Agulhas.
Flowers in October and November.

Oncosiphon suffruticosus
klein-knoppiesstinkkruid

ASTERACEAE

An erect, much-branched annual herb up
to 40cm high. The slightly fleshy leaves
are finely dissected, up to 4cm long, thinly
or densely hairy and pungently aromatic.
Yellow flowerheads, 3–7mm in diameter
and with only yellow disc-florets, are
clustered at the ends of branches; the
hairless involucral bracts are in 4 rows.

    Found in sandy soil, often in disturbed
areas, in Namaqualand, also widespread in
the Western Cape and southern Namibia.
Flowers July to December.

ZELDA WAHL

SCABIOSA
There are nine species of *Scabiosa* in southern Africa
of which one occurs in Namaqualand.

Scabiosa columbaria
jonkmansknoop

DIPSACACEAE

An annual or perennial herb up to
60cm high, branching at the base. The
leaves, crowded at the base of the
plant, are inversely sword-shaped and
entire to deeply lobed. White to mauve
flowerheads up to 2.5cm in diameter are
borne in dense heads at the end of long,
leafless, branched stalks.

    Found in sandy areas in the
Namakwaland Klipkoppe and on the
Knersvlakte, also widespread in the rest of
southern Africa, Europe and Asia.
Flowers August to October.

ZELDA WAHL

ILLUSTRATED GLOSSARY

Figure 1: Leaf arrangements.

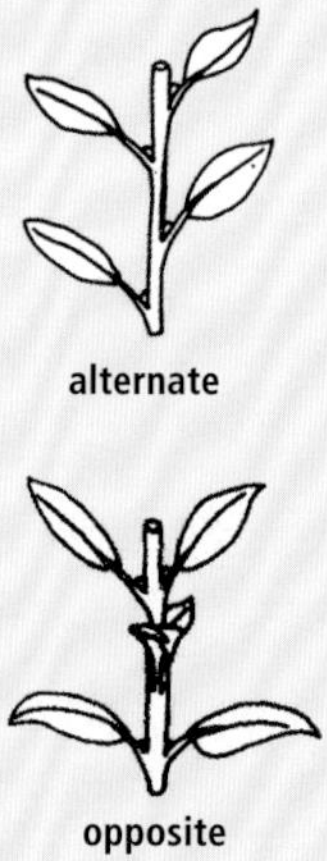

alternate

opposite

Figure 2: A longitudinal section of a typical flower.

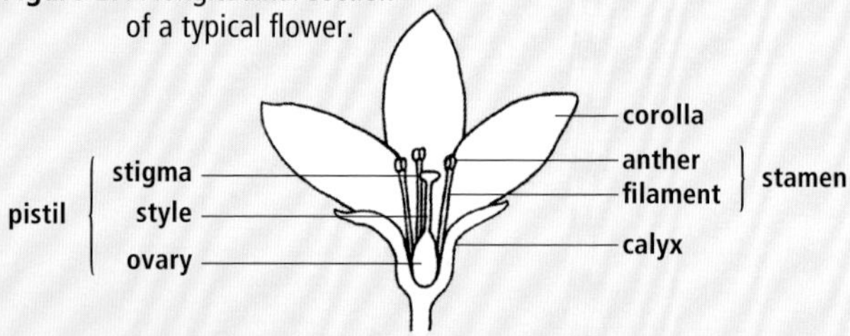

corolla
anther
filament
stamen
stigma
pistil
style
calyx
ovary

Figure 3: A typical leaf.

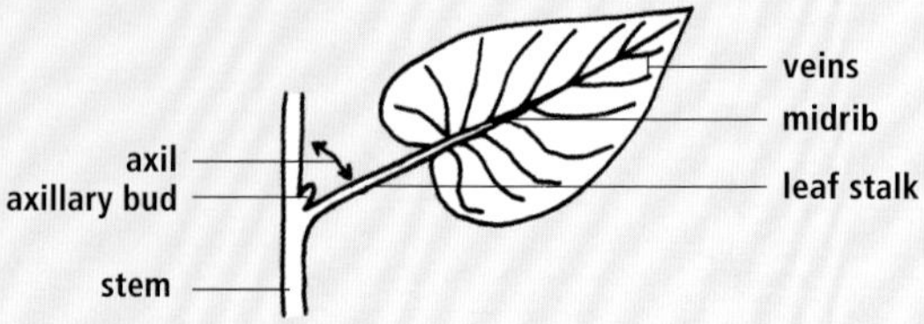

veins
midrib
axil
axillary bud
leaf stalk
stem

Figure 4: Leaf margins.

crisped   entire   scalloped   serrate   sinuate   toothed   undulate

Figure 5: A longitudinal section of a flowerhead of the Asteraceae.

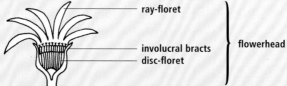

ray-floret
involucral bracts
disc-floret
flowerhead

Figure 6: Leaf shapes.

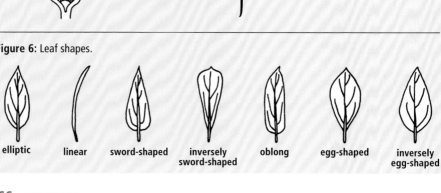

elliptic   linear   sword-shaped   inversely sword-shaped   oblong   egg-shaped   inversely egg-shaped

GLOSSARY OF TERMS

acute: sharply pointed

adpressed: lying closely against the adjacent part, or against the ground

alternate: occurring alternately one above the other, usually pertaining to leaf arrangement on the stem (see Figure 1)

annual: a plant that completes its life cycle in one year or less

anther: the upper pollen-producing portion of a stamen (see Figure 2)

awl-shaped: narrow and gradually tapering to a fine point

axil: the angle between a leaf and the axis that bears it (see Figure 3)

biennial: a plant that completes its life cycle in two years

bract: one of the scale-like or small leaves around a flowerhead or bud or at the base of a flower stalk

bulb: a subterranean storage organ made up of succulent leaf bases

calyx: the outer series of leaf-like segments of a flower, frequently green, sometimes large and colourful, sometimes absent (see Figure 2)

cone: the floral bracts that protect the seeds and that become hardened into a woody cone-shaped head (as in *Leucadendron*, Proteaceae)

corm: a stem that is modified to function as a subterranean storage organ

corolla: the whorl of petals, usually coloured, in a flower (see Figure 2)

crenate: leaf margin notched with regular blunt or rounded teeth

crisped: pertaining to leaf margins that are minutely undulated (see Figure 4)

deciduous: shed at a particular time or season

decumbent: spreading horizontally at first but then growing upwards

disc-floret: a fertile, tubular floret in the central part of the flowerhead in the Asteraceae (see Figure 5)

dissected: deeply cut into many segments

distal: farthest from the point of attachment

dorbank: a distinct underground layer of red soil usually solidified by silica and largely impervious to water. In some places in Namaqualand the topsoil has eroded away and the dorbank is on the surface

doubly lobed: with smaller lobes on the margin of the primary lobes

egg-shaped (ovate): with the outline of an egg, i.e. scarcely twice as long as broad and with the basal half broader than the apical half (see Figure 6)

elliptic: narrowed to rounded at the ends and widest at or about the middle (see Figure 6)

endemic: having a natural distribution confined to a particular geographical region, e.g. found only in Namaqualand and nowhere else in the world

entire: pertaining to a leaf margin without notches or indentations, therefore with an even margin (see Figure 4)

epidermal: of the epidermis, the outermost layer of cells of an organ, usually only one cell thick

evergreen: in leaf throughout the year

ephemeral: a plant with a short life cycle

epidermal: of the epidermis, the outermost layer of cells of an organ,

usually only one cell thick filament: the stalk of an anther (see Figure 2)

filamentous staminodes: filaments without a perfect anther, such stamens are therefore infertile

flowerhead: a close head of flowers, usually pertaining to the Asteraceae (see Figure 5), or many flowers borne very close together as in the Proteaceae

fluted: with channels or grooves

herb: a plant with no woody parts above ground

internode: the portion of a stem between two nodes (compare node)

inversely egg-shaped (obovate): with the outline of an egg, i.e. scarcely twice as long as broad but with the apical half broader than the basal half (see Figure 6)

inversely sword-shaped (oblanceolate): tapering towards the base more than towards the tip (see Figure 6)

involucral bracts: one or more whorls of bracts or small leaves surrounding the base of a flowerhead as in the Asteraceae (see Figure 5)

lignified: to become woody through the formation and deposit of lignin in cell walls

linear: long and narrow, the sides parallel or nearly so (see Figure 6)

lobe: a part of a petal, sepal or leaf that represents a division to about halfway to the centre

midrib: the principal vein in the centre of a leaf that is a continuation of the leaf stalk (see Figure 3)

node: a point on the stem at which a leaf or leaves and accompanying organs are borne (compare internode)

oblique: with unequal sides, e.g. applied to a leaf with the two sides of the blade unequal at the base

oblong: longer than broad, the sides more or less parallel for most of their length (see Figure 6)

opposite: borne two at a node, on opposite sides of the stem (see Figure 1)

orbicular: flat with a more or less circular outline

ovoid: solid shape with an egg-shaped outline

papilla (pl. papillae): a small protuberance

pedicel: the stalk of one flower in a cluster

perennial: a plant that lives for more than two years

petal: a unit of the corolla of a flower, usually coloured and more or less showy – in this guide also referring to a tepal (i.e. any member of the outer sterile floral envelopes that are not clearly differentiated into petals and sepals)

prostrate: lying flat on the ground

ray-floret: a ribbon-like flower, usually sterile, on the outer margins of the flowerhead in the Asteraceae (see Figure 5)

recurved: rolled or curved backwards

rhizome: a subterranean stem forming roots and usually producing leaves at the tip

scale: a kind of small, mostly dry and appressed leaf or bract

scalloped: regularly indented, like some leaf margins (see Figure 4)

sepal: a unit of the calyx of a flower, usually green but can be coloured, as in the Orchidaceae

serrate: margin with saw-like teeth (see Figure 4)

shrub: a plant with multiple woody stems branching from at or near the ground. In Namaqualand these shrubs can be anything from 5cm to 1m high, but mostly 25–75cm high

sinuate: with the margin strongly wavy (see Figure 4)

spathulate: broad at the apex and tapered to the base

spur: an elongated tubular pouch at the base of a petal

stamen: the pollen-producing male part of a flower, consisting of an anther and a filament (see Figure 2)

staminode: an abortive stamen without a perfect anther

stipule: a leaf-like or scale-like appendage at the base of a leaf stalk

stolon: a slender, prostrate or trailing stem that produces roots

sword-shaped (lanceolate): longer than broad and tapering towards the tip (see Figure 6)

tendril: a long, slender, coiling, modified leaf, or rarely stem, by which a climbing plant attaches to its support

toothed: with sharp projections on a margin (see Figure 4)

truncate: appearing as if cut off at the end, nearly or quite straight across

tuber: a root or stem that is modified to function as a subterranean storage organ

tubercule: a small swelling or outgrowth

undulate: with a wavy margin (see Figure 4)

valves: flap-like coverings in the dry fruit of the vygie group in the Aizoaceae that open when the fruit is wet and close again when it dries out

water-storing epidermal cells: a single layer of large, conspicuous cells on a leaf or stem surface in which water is stored

INDEX

Accepted scientific names are in **bold italic**; synonyms are in *italic*; family names are in CAPITALS.

Q

Quaqua acutiloba (N.E.Br.) Bruyns 344
Q. framesii (Pillans) Bruyns 344
Q. incarnata (L.f.) Bruyns 345
Q. mammillaris (L.) Bruyns 345
quiver tree 98

R

raaptol 41
Radyera urens (L.f.) Bullock 250
ramblom 130
ramhorinkies 367
rankbrakbossie 267
rank-t'nouroe 308, 333
rankvy 307
rankvygie 333
rapuis 421
rasmusbas 185
reksening 428
renosterbos 441
RESTIONACEAE 143
RHAMNACEAE 212
Rhus horrida, see **Searsia horrida** 240
R. incisa, see **Searsia incisa** 240
R. longispina, see **Searsia longispina** 241
R. populifolia, see **Searsia populifolia** 241
R. undulata, see **Searsia undulata** 242
Rhynchopsidium pumilum (L.f.) DC. 437
Rhynchosia schlechteri Baker f. 206
ribbokboegoe 244
Ricinus communis L. 226
ridderspoor 235
river lily 106
rivierlelie 106
rock chinkerinchee 126
Roepera cordifolia (L.f.) Beier & Thulin 186
R. foetida (Schrad. & J.C.Wendl.) Beier & Thulin 186
R. leptopetala (E.Mey. ex Sond.) Beier & Thulin 187
R. morgsana (L.) Beier & Thulin 187
R. sp. 188
R. teretifolia (Schltr.) Beier & Thulin 188
Rogeria longiflora (Royen) J.Gay ex DC. 381
rolbos 263
Romulea citrina Baker 56
R. kamisensis M.P.de Vos 56
R. minutiflora Klatt 57
R. namaquensis M.P.de Vos 57
rooi-dirkie 125
rooidoring 240

rooikalkoentjie 49
rooikanol 141
rooipootjie 144
rooipoppies 387
rooisaadgras 144
rooisuring 253
rooi-t'kooi 335
rooiviooltjie 137
ROSACEAE 211
Rumex cordatus Poir. 253
RUSCACEAE 116–117
Ruschia aggregata L.Bolus 332
R. muelleri (L.Bolus) Schwantes 332
R. robusta L.Bolus 333
R. viridifolia L.Bolus 333
RUTACEAE 244–245

S

salie 384, 385
Salsola aphylla, see **Caroxylon aphyllum** 263
S. kali, see **Kali** sp. 263
Salvia africana-lutea L. 384
S. dentata Aiton 384
S. lanceolata Lam. 385
S. verbenaca L. 385
sambreelblom 106
sambreeltjies 446, 447
sandbietou 429
sandkalossie 137
sandklawer 202
sandlelie 62, 129
sandpampoentjie 250
sandsalie 384, 385
sandslaai 299–301
SANTALACEAE 255–256
SAPINDACEAE 242–243
Sarcocaulon ciliatum, see **Monsonia ciliata** 172
S. crassicaule, see **Monsonia crassicaulis** 173
S. flavescens, see **Monsonia flavescens** 173
S. herrei, see **Monsonia herrei** 174
S. l'heritieri, see **Monsonia spinosa** 176
S. multifidum, see **Monsonia multifida** 174
S. patersonii, see **Monsonia patersonii** 175
S. salmoniflorum, see **Monsonia salmoniflora** 176
Sarcocornia littorea (Moss) A.J.Scott 264
S. natalensis (Bunge ex Ung.-Sternb.) A.J.Scott subsp. **affinis** (Moss) S.Steffen, Mucina & G.Kadereit 264
S. pillansii (Moss) A.J.Scott 265
S. xerophila (Toelken) A.J.Scott 265

Sarcostemma viminale (L.) R.Br. 351
Satyrium erectum Sw. 36
S. pumilum Thunb. 36
Scabiosa columbaria L. 465
Sceletium exalatum, see **Mesembryanthemum exalatum** 287
scented turpentine grass 152
Schismus schismoides (Stapf ex Conert) Verboom & H. P. Linder 150
Schmidtia kalahariensis Stent 147
SCROPHULARIACEAE 366–380
sea coral 264
sea heath 254
sea pumpkin 403
sea saltbush 267
Searsia horrida (Eckl. & Zeyh.) Moffett 240
S. incisa (L.f.) F.A.Barkley var. **effusa** (C.Presl) Moffett 240
S. longispina (Eckl. & Zeyh.) Moffett 241
S. populifolia (E.Mey. ex Sond.) Moffett 241
S. undulata (Jacq.) T.S.Yi, A.J.Mill. & J.Wen 242
seebiesie 149
seebrakbos 267
seed of Asteraceae genera 400–401
seekoraal 264
seepampoen 403
seepbos 266
seervingerplakkie 162
Selago albida, see **Selago namaquensis** 379
S. glabrata Choisy 378
S. glutinosa E.Mey. 378
S. namaquensis Schltr. 379
Senecio arenarius Thunb. 416
S. bulbinifolius DC. 416
S. cardaminifolius DC. 417
S. cicatricosus Sch.Bip. 417
S. cinerascens Aiton 418
S. longiflorus (DC.) Sch.Bip. 418
S. radicans (L.f.) Sch.Bip. 419
S. sisymbriifolius DC. 419
sentkannetjie 163
Septulina glauca (Thunb.) Tiegh. 257
Sesamum capense Burm.f. 381
Sesuvium sesuvioides (Fenzl) Verdc. 278
shepherd's tree 232
sifkopkandelaar 158
Silene undulata Aiton 260
Sisyndite spartea E.Mey. ex Sond. 192
sjambokbossie 418
sjielingbos 186
skaapbos 432